CISM Courses and Lectures

The series presents lecture notes, monographs, edited works and proceedings in the field of Mechanics, Engineering, Computer Science and Applied Mathematics.
Purpose of the series is to make known in the international scientific and technical community results obtained in some of the activities organized by CISM, the International Centre for Mechanical Sciences.

International Centre for Mechanical Sciences

Courses and Lectures Vol. 546

For further volumes:
www.springer.com/series/76

Markus J. Buehler and Roberto Ballarini
Editors

Materiomics: Multiscale Mechanics of Biological Materials and Structures

Springer

Editors

Markus J. Buehler
Civil and Environmental Department, Massachusetts Institute of Technology,
Cambridge, MA, USA

Roberto Ballarini
Department of Civil Engineering, University of Minnesota, Minneapolis, USA

ISSN 0254-1971
ISBN 978-3-7091-1573-2   ISBN 978-3-7091-1574-9 (eBook)
DOI 10.1007/978-3-7091-1574-9
Springer Wien Heidelberg New York Dordrecht London

All contributions have been typeset by the authors
Printed in Italy

Printed on acid-free paper

Springer is part of Springer Science+Business Media (www.springer.com)

PREFACE

This book contains lecture notes from leading researchers in the field of mechanical sciences of biological materials and structures, with a focus on the behavior of biological materials under extreme physical, chemical, physiological and disease conditions, as well as on biomimetic and bioinspired material development for technological applications. To provide a thorough foundation for this research, the course will focus on the integration of advanced experimental, computational and theoretical methods applied to the study of biological materials across disparate length- and time-scales. Specific attention is paid to the integration of theoretical, computational and experimental tools that could be used to assess structure-process-property relations and to monitor and predict mechanisms associated with the function and failure of biological materials and structures composed of them.

The chapters provide overviews of emerging fields of research and highlight important challenges and opportunities. Hence, the three core objectives of this book are to: (1) Provide a clear description of methods and tools, (2) Present case studies that demonstrate the impact of multiscale modeling approaches, and to (3) Provide a carefully selected list of core references and citations for the interested reader. The case studies include the analysis of key biological materials, the biodegradation of implanted synthetics, the transfer of biological material principles towards bioinspired applications, and the exploration of diseases in which material failure plays a critical role. The approaches presented in this book emphasize the fundamental principles of physics, chemistry and mechanics, and they rely on quantum mechanics, molecular dynamics and continuum analyses. The use of basic sciences creates a powerful common platform regardless of the specific material system considered, and can therefore be transferred to other types of materials and structures.

The editors of this volume would like to thank the CISM team for their help and support in preparing this book. They are also grateful to the contributors of the various chapters for their time and efforts, and acknowledge the support of their research on the mechanical behavior of materials and structures over the years from the National Science Foundation, the Army Research Office, the Office of Naval

Research, DARPA, the Air Force Office for Scientific Research, and the National Institutes of Health.

Roberto Ballarini and Markus Buehler

CONTENTS

Introduction

Roberto Ballarini * and Markus Buehler †

* Department of Civil Engineering, University of Minnesota, Minneapolis, USA
† Department of Civil and Environmental Engineering, Massachusetts Institute of Technology, Cambridge, USA

1 The promise of multiscale modeling and bioinspired engineering

The field of multiscale mechanics has witnessed an exciting development over the past decades, culminating in recent years in breakthrough discoveries that have blurred the boundaries between living and synthetic materials, and have enabled the first wave of high-impact applications of new materials and structures in biomedical, energy and structural engineering applications. Multiscale modeling offers promise for facilitating the creation of engineered materials and structures with properties that resemble those of biological systems, in particular the ability to self-assemble, to self-repair, to adapt and evolve, and to provide multiple functions that can be controlled through external cues. In addition to their potential for enabling the realization of advanced technological applications, the challenges posed by the complex behavior of hierarchical tissues and cells in biological systems represent terrific opportunities to open new chapters in the development of the mechanical sciences. It is remarkable how the mechanics practiced by da Vinci, Galileo, Newton and other great scientists has evolved to a point where now it interconnects intimately with the life sciences, and that it could ultimately contribute to the solutions of critical problems encountered in such disparate fields as medicine and the aging infrastructure.

Yet, in spite of significant advancements in the study of biological materials in the past decade, a lack of sufficient understanding of the fundamental physics of many phenomena in biology is hindering our progress towards the building of sufficiently robust models, simulation tools and experimentation. For example, the understanding of the mechanisms of failure in biological systems remains elusive, including those involved in the breakdown of diseased tissue, the failure of biological components due to injuries, and the ability of biological systems to mitigate adverse effects of damage through self-healing mechanisms. The cost-effective manufacturing of bioinspired products is also an enormous challenge, because humans have traditionally relied on top-down fabrication paradigms that simply cannot be used to efficiently produce the highly hierarchical structures that *Nature* builds from the bottom-up. Improved understanding of how biology originates from the molecular scale and proceeds to genes (DNA), proteins, tissues, organs and organisms can guide our development of self-assembly technologies that will allow mass production and utilization of bioinspired materials for daily life applications like consumer products, medical devices and large-scale systems in the aerospace, defense and building industries.

The highly complex nature of biological structures, which involve multiphysics and multiple length and time scales, has inspired the new field of study referred to as materiomics, which is defined in the next section and is reflected by the contents of this book.

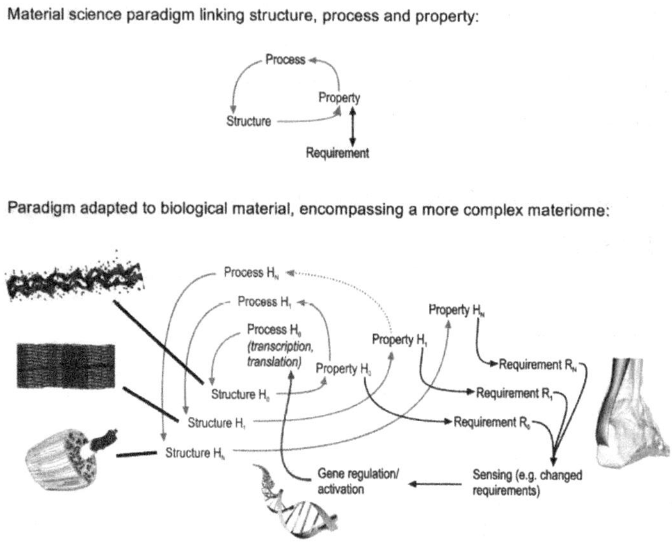

Figure 1. Schematic representing the materiomics approach (Cranford and Buehler (2012)).

2 Materiomics

What is materiomics? As illustrated schematically in Figure 1, it is an approach rooted in physics that extends the structure-process-property-requirement paradigm that has been developed by materials scientists to the analysis of highly complex biological and synthetic materials and structures. The schematic emphasizes that materiomics is a holistic systems approach to the theoretical, computational and experimental study of materials that aims to identify links between processes, structures, and properties across multiple scales, from nano to macro. The integrated view and description of the building blocks of a hierarchical structure and their fundamental interactions is referred to as the material's materiome. Materiomics thus provides a systematic description of universal mechanisms by which complex system functionality and failure can be explained from the materiome. As sketched in Figure 2, similar to the way music is created from a finite number of musical notes, the relationships between form and function found in natural materials provide the mechanistic basis to explain the remarkable mechanical properties of materials like nacre, bone, spider silk and collagen. For example, it has been determined that the toughness of bone and of sea shells of the crossed-lamellar type are the result of multiple and synergetic toughening mechanisms made possible by a half-dozen distinct microstructural features (Kamat et al. (2000); Ballarini et al. (2005)). In fact, for bone, sea shells and most other biological objects the traditional concepts of "structure" and "material" are blurred, and integrated in a vision that derives functional properties by systematically and strategically adapting multiple levels across numerous length- and time-scales (Figure 2). This viewpoint extends our current ability to engineer structures to the desired scale, and requires a multidisciplinary treatment of problems to incorporate physics, chemistry and advanced mathematics to develop complex models to design and predict performance.

New approaches that take advantage of mathematical tools such as material ologs (Spivak et al. (2011)) are important in arriving at a systematic analysis that reduces complexity to distill the essential features. Moreover, a range of experimental and computational tools is needed to measure and produce structures and properties at these variegated length- and time scales. A summary of key computational methods, synthesis, processing and experimental techniques is provided in Figure 3. It is evident that with modern tools a very broad range of scales can be seamlessly explored, thus allowing the realization of realistic multiscale analysis. Figure 4 depicts an impressive example of a precise experimental analysis of collagen microfibrils using the microtechnology-based material testing described in Chapter 3, something that would have been impossible just a few years ago.

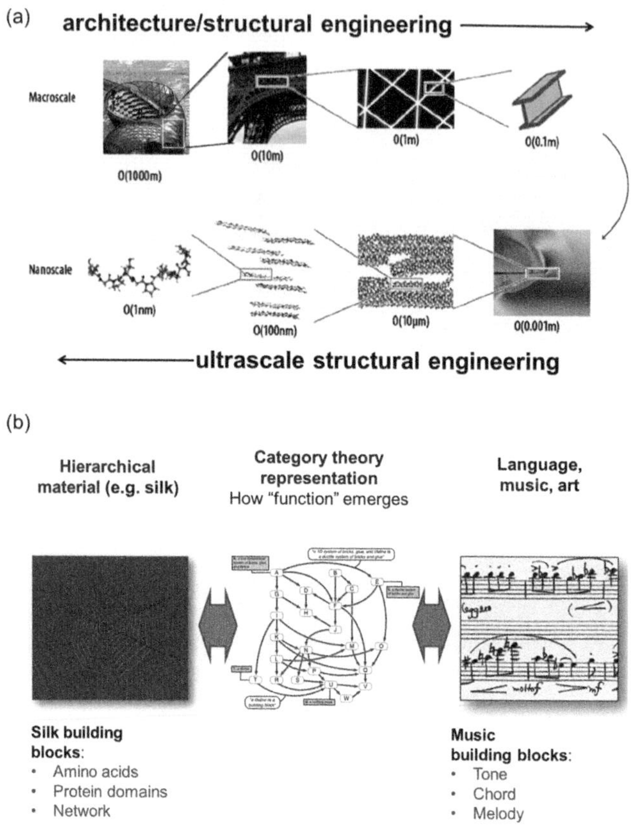

Figure 2. Common principles of biological and bioinspired material design, showing the merger of "structure" and "material" across different length scales in hierarchical materials (Buehler and Ackbarow (2007)).

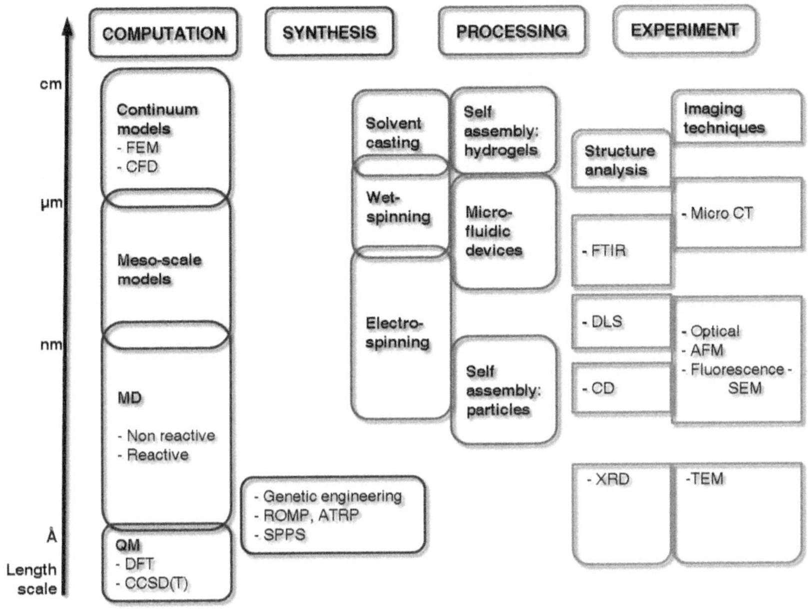

Figure 3. Overview of various computational and experimental tools, including synthesis, processing and imaging/manipulation techniques. The emergence of tools operating at different length scales now enables the analysis of materials across all relevant scales (Gronau et al. (2012)).

3 Motivation for studying biological structures: The superior performance of natural structures conferred by their hierarchical designs

Why should humans study biological structures and paradigms? Because *Nature* has created an extremely large number of high-performance prototypes that humans can reverse engineer and in turn use as inspirations for creating synthetic products with similar superior performances. This section focuses on but one strategy used by *Nature* to create materials and structures whose survival requires superior mechanical properties and structural behavior, namely highly hierarchical design.

The structures created by organisms, although made from rather mundane materials, show impressive properties that are clearly well-suited for

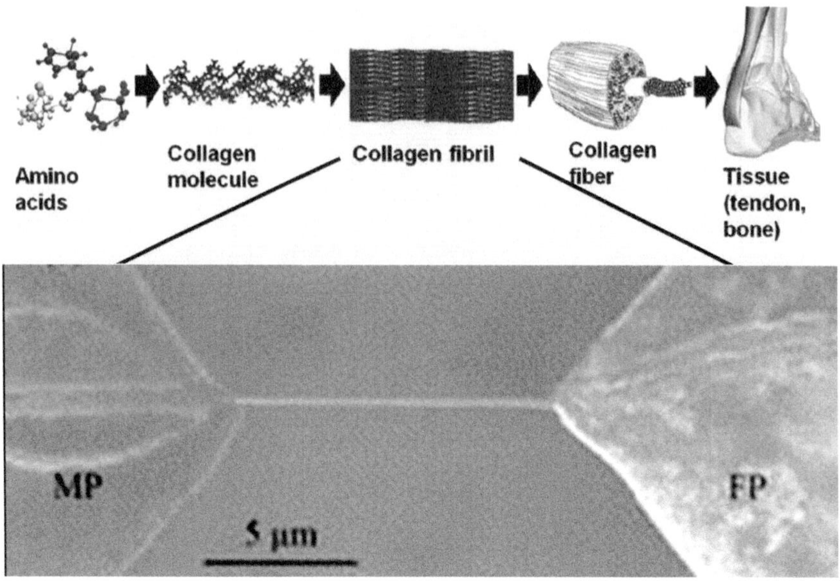

Figure 4. Example of advanced experimentation applied to test the nanomechanics of individual collagen fibrils. Results from such experiments can be compared to molecular modeling and enable us to ask fundamental questions about the physiology and disease of key construction materials in nature (adapted from Eppell et al. (2006)).

their intended functions. Structure/function/performance correlations have been assumed by scientists to be a result of evolutionary pressures inherent in natural selection. While proponents of "intelligent design" offer other explanations, the diversity of microstructure in structures such as molluscan shells and bone and their remarkable mechanical properties and self-healing mechanisms testifies to the flexibility and power of this approach. *Nature* achieves robust structures using as little mass as possible, through judicious arrangements of mundane polymeric and ceramic components. Be they primarily ceramic (tooth enamel, mollusc shell), polymeric (insect exoskeleton, plant cell walls), or more evenly balanced composites (antler, bone), biological materials are virtually all composites utilizing different proportions of the basic components and a variety of hierarchical structural architectures.

It is very instructive to compare the mechanical properties of biological structures with materials created by humans through the performance index paradigm pioneered by Ashby (Ashby (1992)). First, a brief primer on basic structural mechanics. Consider the simple extension of a rod of length, L, and cross-sectional area, A, made of a material with density, ρ (Figure 5). In terms of the applied force, F, and the elongation, δ , the stress and strain are defined as $\sigma = \frac{F}{A}$ and $\varepsilon = \frac{\delta}{L}$. For most engineering materials, the relation between σ and ε is linear at small values of strain, with a slope defined as the elastic (Young's) modulus, E. At the elastic limit, σ_f, the curve ends abruptly for brittle materials and is nonlinear for ductile materials up to the ultimate stress required to fracture the rod, σ_u. The area under the linear part of the curve up to a given strain, $\frac{\sigma^2}{2E}$, is defined as the elastic strain energy density and represents the potential energy conferred to the rod by the work performed by the applied force. If the strain is limited to values less than the yield strain, $\varepsilon_f = \frac{\sigma_f}{E}$, then the tie will return to its original length upon removal of the force. This behaviour is referred to as elastic, in that no energy is dissipated during a loading-unloading cycle. The total area under the $\sigma - \varepsilon$ curve represents the work done by the force to fracture the rod into two pieces; the work of fracture is defined as this work divided by the area of the surfaces created by fracturing the rod into two pieces.

If a relatively brittle structure contains a crack-like flaw and is treated as linear elastic, the stresses along the crack front are singular and therefore cannot be directly used to predict load carrying capacity. Instead, the force required to fracture the structure is determined by the stress intensity factor, K, which characterizes the stress and strain intensities in the vicinity of the crack front. The stress intensity factor depends on the geometry of the structure, the type of loading, and the specific crack shape; according to linear elastic fracture mechanics theory the crack will extend across the specimen when K reaches a critical value defined as the fracture toughness, K_c. The stress intensity factor is directly related to J, the energy available to overcome the material's resistance to crack extension, by the equation $K = \sqrt{EJ}$. Therefore the fracture toughness can be expressed in terms of the energy required to create the fracture surfaces, J_c, by the relation $J_c = \frac{K_c^2}{E}$.

Quantitative comparisons between materials can be made using the concept of material performance indices, parameters that quantify a material's ability to perform a certain function. The higher the value of the index, the better suited is the material for a given application. For a thorough discussion of the mechanical properties of natural materials and the origins of their superiority, the reader is referred to Ashby et al. (1995) and Wegst and Ashby (2004). Here we borrow from their discussions.

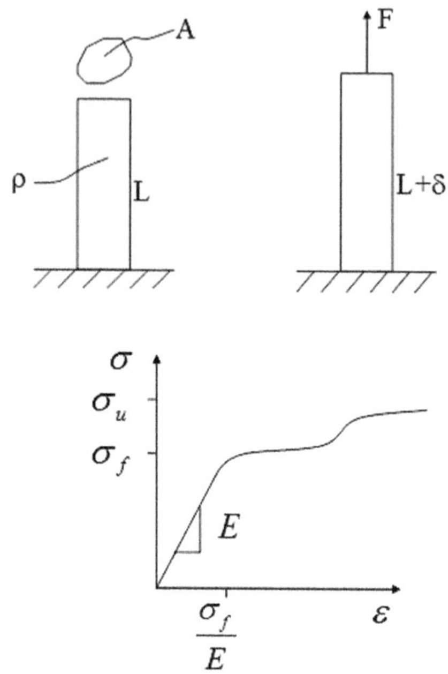

Figure 5. Simple elongation experiment and representative stress-strain curve.

Table 1. Performance indices for different types of structural elements.

Design	Tie in tension	Beam in flexure	Plate in flexure
Maximum strength to weight	$\frac{\sigma_f}{\rho}$	$\frac{\sigma_f^{2/3}}{\rho}$	$\frac{\sigma_f^{1/2}}{\rho}$
Maximum stiffness to weight	$\frac{E}{\rho}$	$\frac{E^{1/2}}{\rho}$	$\frac{E^{1/3}}{\rho}$
Large recoverable deformation	$\frac{\sigma_f}{E}$	$\frac{\sigma_f}{E}$	$\frac{\sigma_f}{E}$
Spring with minimum volume	$\frac{\sigma_f^2}{E}$	$\frac{\sigma_f^2}{E}$	$\frac{\sigma_f^2}{E}$
Fracture safe displacement controlled design	$\left(\frac{J_c}{E}\right)^{\frac{1}{2}}$	$\left(\frac{J_c}{E}\right)^{\frac{1}{2}}$	$\left(\frac{J_c}{E}\right)^{\frac{1}{2}}$

Consider first the minimum weight design of three simply loaded structures; a rod in tension, a flexed beam, and a flexed plate. The first two rows of Table 1 list the performance indices relevant to a minimum weight design of these structures that does not specify the shape of the element's cross-section. The fact that the most efficient material for a strong and light tie (or similarly stiff and light) subjected to simple elongation is the one that possesses the largest value of $\frac{\sigma_f}{\rho}$ (or similarly $\frac{E}{\rho}$) is perhaps obvious and can be guessed by someone that has never taken a course in mechanics. However, those that have studied structural mechanics know that beam and plates experience spatially varying stress and strain distributions that lead to performance indices that involve different rational exponents to the strength and stiffness.

Table 2. Stiffness to weight performance indices. The parentheses reflect the elastic modulus along the stiffer direction, and therefore are not representative of a stiff plate in bending that requires equal moduli in all directions (adapted from Ashby et al. (1995); Wegst and Ashby (2004)).

Material	E/ρ $(GPa/Mg/m^3)$	$E^{\frac{1}{2}}/\rho$ $(GPa^{\frac{1}{2}}/Mg/m^3)$	$E^{\frac{1}{3}}/\rho(GPa^{\frac{1}{3}}/Mg/m^3)$
Palm (Iriartea)	23	12.5	(10.1)
Mild steel	27	1.8	(0.8)
Balsa wood	20	14.1	(12.6)

Consider specific materials. Table 2 shows that for the stiff and light design, the performance index of mild steel is comparable (actually slightly higher) to that of the two representative woods. However, steel is not nearly as efficient as Palm or Balsa for beams or plate elements subjected to flexure. This data suggests that the various hierarchical microstructures of wood have evolved to ensure that trees are efficient in their response to the principal loads they must carry; bending of branches under their own weight and bending of trunks under wind loads. Similar insights are provided by Table 3 for the strong and light design criterion.

We now turn our attention to material choices for an elastic hinge, a component that is required to undergo relatively large deformation when loaded, and to return to its original shape when the load is removed. The performance index for the design of elastic hinges is $\frac{\sigma_f}{E}$, and it as listed in Table 4, leather, cartilage, and to a lesser extent, skin, are the best choices. To man's credit, rubber edges out its biological counterparts, although rubber is not usually thought of as a bioinspired material.

Table 4 also shows that silk fibre is the top choice for springs that are required to absorb the most energy per unit volume. In fact, it out performs

Table 3. Strength to weight performance indices.The parentheses reflect the elastic modulus along the stiffer direction, and therefore are not representative of a stiff plate in bending that requires equal moduli in all directions (adapted from Ashby et al. (1995); Wegst and Ashby (2004)).

Material	σ_f/ρ $(GPa/Mg/m^3)$	$\sigma_f^{\frac{2}{3}}/\rho$ $(GPa^{\frac{2}{3}}/Mg/m^3)$	$\sigma_f^{\frac{1}{2}}$ $(GPa^{\frac{1}{2}}/Mg/m^3)$
Mild steel	51	6.9	2.5
Balsa wood	160	64	(40.0)

man-made spring steel , and the natural elastic hinges leather and cartilage by significant margins. Its molecular design guarantees that silk can absorb, without fracturing, the energy introduced by the excursions of the spider and/or the desperate movements of its entangled prey.

Table 4. Performance indices relevant to hinges and springs (adapted from Ashby et al. (1995); Wegst and Ashby (2004)).

Material	σ_f/E	σ_f^2/E (MJ/m^3)
Spring steel	0.01	19.0
Soft butyl rubber	1.40	19.6
Single silk fiber	0.14	290.0
Cartilage	1.00	10.0
Skin	0.25	2.5
Leather	1.00	45.0

The last example involves cracked structures. Consider the specific case of choosing a material for a cracked structure that is required to survive large elastic deformation. Table 5 indicates that skin has the best performance index, despite having a much lower fracture toughness than steel or highly mineralized mollusc shell. Interestingly, the microstructural design of skin may have evolved to make it ideally suited for being stretched around our knuckles and elbows without tearing. We could not enjoy such movements if our skin had the fracture properties of man-made alloys or other biological materials.

The biological materials just described can therefore be considered prototypes of successful structural designs that *Nature* has provided to gratis and that can inspire us to create our own tailor made materials with similar performances. The major challenge is the development of methods for fabricating such highly hierarchical structures.

Nature is indeed impressive and inspiring.

Table 5. Performance indices for structures containing cracks (adapted from Ashby et al. (1995); Wegst and Ashby (2004)).

Material	$K_c = (EJ_c)^{\frac{1}{2}}$ $(MPa - m^{\frac{1}{2}})$	$(\frac{J_c}{E})^{\frac{1}{2}}$ $(mm^{\frac{1}{2}})$
Mollusc shell	9.5	0.4
Mild steel	90.0	0.4
Skin	0.4	38.7

Bibliography

M.F. Ashby. *Materials selection in mechanical design.* Pergamon Press, 1992.

M.F. Ashby, L.J. Gibson, U. Wegst, and R. Olive. The mechanical properties of natural materials. i. material property charts. *Proc. R. Soc. Lond. A*, 450:123–140, 1995.

R. Ballarini, R. Kayacan, F.J. Ulm, T. Belytschko, and A.H. Heuer. Biological structures mitigate catastrophic failure through various strategies. *International Journal of Fracture*, 135:187–197, 2005.

M.J. Buehler and T. Ackbarow. Fracture mechanics of protein materials. *Materials Today*, 10(9):46–58, 2007.

S. Cranford and M.J. Buehler. *Biomateriomics.* Springer, 2012.

S. Eppell, B. Smith, H. Kahn, and R. Ballarini. Anano measurements with micro devices: Mechanical properties of hydrated collagen fibrils. *Journal of the Royal Society Interface*, 3:117–121, 2006.

G. Gronau, S. Tarakkad Krishnaji, M.E. Kinahan, T. Giesa, J.Y. Wong, D.L. Kaplan, and M.J. Buehler. A review of biopolymer structure-process-property relationships at multiple scales via integration of modeling and experimentation. *Biomaterials*, accepted for publication, 2012.

S. Kamat, X. Su, R. Ballarini, and A.H. Heuer. Structural basis for the fracture toughness of the shell of the conch strombus gigas. *Nature*, 405: 1036–1040, 2000.

D. Spivak, T. Giesa, E. Wood, and M.J. Buehler. Fracture mechanics of protein materials. *PLoS ONE*, 6(9):e23911, 2011.

U.G.K. Wegst and M.F. Ashby. The mechanical efficiency of natural materials. *Philosophical Magazine*, 21:2167–2181, 2004.

Multi-scale modeling of biomaterials and tissues

Alfonso Gautieri[1, 2] Markus J. Buehler[1, *]

[1] Civil and Environmental Department, Massachusetts Institute of Technology, Cambridge, MA, USA

[2] Department of Bioengineering, Politecnico di Milano, Italy

Abstract

Computer simulation has emerged as a powerful tool to investigate and design materials without ever making them. Predicting the properties and behavior of materials by computer simulation from the bottom-up perspective has long been a vision of computational materials scientists and, as computational power increases, modeling and simulation tools are becoming crucial to the investigation of material systems. The key to achieving this goal is using hierarchies of paradigms that seamlessly connect quantum mechanics to macroscopic systems. Particular progress has been made in relating molecular-scale chemistry to mesoscopic and macroscopic material properties essential to define the *materiome*. This chapter reviews large-scale atomistic and coarse-grain modeling methods commonly implemented to investigate the properties and behavior of natural and biological materials with nanostructured hierarchies. We present basic concepts of hierarchical multiscale modeling capable of providing a bottom-up description of chemically complex materials and some example applications related to the study of collagen material at different hierarchical levels.

1 Introduction to the multi-scale modeling paradigm

In recent years, researchers consider computer simulations as a tool to design new materials, new structures, or to develop new drugs, without the need to synthesize them. A vision of materials scientists has long been to

* Corresponding author at: Laboratory for Atomistic and Molecular Mechanics, Department of Civil and Environmental Engineering, Massachusetts Institute of Technology, 77 Massachusetts Ave., Room 1-235A&B, Cambridge, MA, USA.Tel.: +1-617-452-2750. E-mail address: mbuehler@MIT.EDU (M.J. Buehler).

predict the properties and behavior of materials by computer simulation from a fundamental or *ab initio* perspective. With the increasing computational power, modeling and simulation tools become indispensable to the investigation of material systems. The key to achieving this aim is utilizing hierarchies of paradigms and scales that seamlessly connect macroscale systems to first principal quantum mechanics. Particular progress has been made in relating atomistic-scale chemistry to mesoscale and macroscale material properties essential to define the *materiome* (see Fig. 1). This chapter will review large-scale atomistic and molecular modeling methods commonly implemented to investigate the properties and behavior of natural and biological materials with nanostructured hierarchies. We present basic concepts of hierarchical multiscale modeling able of providing a bottom-up description of chemically complex materials.

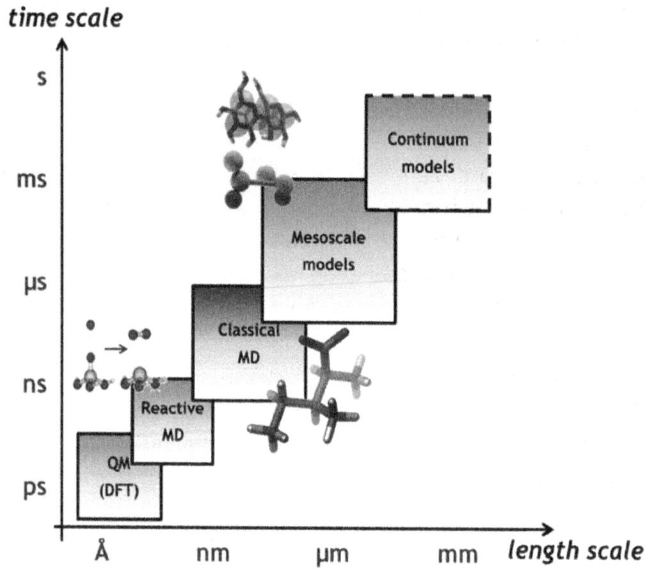

Figure 1. Hierarchy of characteristic/accessible time and length scales for computational materiomics.

Multiscale simulation techniques for biological materials have become increasingly popular in recent years and have enabled the direct link between theoretical description and experimental characterization of materials. These methods are necessary when a single level of resolution is inadequate to probe complex interactions that act between scales, and when hierarchies are intimately interwoven. Even if many questions about material systems can be studied within a single-scale and with a single level of resolution, the use of theoretical and computational approaches enables critical progress in linking the chemical or molecular, and mesoscale structures of these materials to macroscale properties, across disparate scales (Cranford and Buehler, 2012). Such studies, for example, are crucial in understanding the impact of hierarchical structures, genetic mutations, structural flaws and defects, environmental variations (such as temperature, solvent or pH changes), and other chemical stimuli on the properties of materials. The combined use of a set of computational methods is crucial to cover all relevant material properties and scales. Such approach includes first principles calculations, atomistic modeling, coarse-grained models, as well as continuum theory based methods. The properties of biological materials (*e.g.*, proteins) are determined by interactions on a wide range of length- and time-scales. As there is intrinsically mutual influence between scales (scale-bridging effects), it is difficult to provide quantitative information and understanding at a single scale (whether at the ultimate functional level or some intermediate regime) without taking this properly into account. Nevertheless, in the field of atomistic-based multiscale modelling, it is now possible to start from the bottom scales (considering electrons and atoms), to reach all the way up to macroscale scales of filaments, fibrils, fibers and entire tissues, by explicitly considering the characteristic structural features at multiple materiomic hierarchies. Such approaches are possible with the advent of first principles based multiscale simulation techniques. The fundamental principle underlying these multiscale simulation methods is *finer scales train coarser scales* a bold approach with enormous potential to change the way we investigate and design materials (Fig. 1). This progress now provides one with many opportunities and, even though there are still major challenges, this approach could transform the materials science of biological and other materials as a discipline through increased integration of computational methods in scientific research. Despite significant advances in the study of biomolecules and biological materials, the fundamental physics of many biological phenomena continue to pose a substantial challenge for modeling and simulation. One strategy is to devise independent models, which can describe the behavior of the material at a given scale. Then, a collection of single-scale descriptions across scales can provide a practical (and theoretically sound) description of the material. The parameters characterizing each scale independently provide the link between scales. A good illustration of this approach is the separation of quantum effects, dislocation theory, and continuum models of metals. This strategy is quite successful when it comes to the description of materials and it is used in many applications. A quantum-level approach may be utilized for nanoelectronics, for example, whereas a continuum perspective for larger mechanical components. In this way, the *application scale* determines the *model resolution*. However, even though this strategy delineates the material properties (and thus it is suitable for a given application), it gives no insight into the *materiome*, nor it gives indication of how changes at one scale can affect another scale. It is the contention of materiomics that complex hierarchical material systems, by definition, prohibit such separation of scalesthey are intrinsically woven. A continuum model cannot be correctly formulated without considering molecular effects, and molecular effects can be highly dependent on quantum level pehnomena. Thus, a complete picture of the material system require incorporating all scalesthe materiome. That said, from a materiomic perspective, a multiscale methodology is not only desired but necessary, in order to not only describe material properties but to understand the structural organization and physical mechanisms at different scales that together contribute to material function.

2 Classical molecular dynamics

An effective instrument in the investigation of any materiome is molecular modeling. The continual advances in computational power have made molecular modeling approaches more accessible (and powerful) than ever before. However, the theoretical foundations of molecular modeling can be traced back to the mid-20th century. Full atomistic Monte Carlo methods can be attributed slightly earlier to the work of Metropolis, Ulam, Teller and co-workers (Metropolis et al., 1953), while the molecular dynamics (MD) method was first announced by Alder and Wainwright in the late 1950s (Alder and Wainwright, 1960, 1959, 1957) to study the interactions of hard spheres mimicking ideal atoms. Even with such simple engineering approximation, many important insights concerning the behavior of simple liquids emerged from their pioneering studies. Then, in 1964 the first molecular dynamics simulations using a realistic potential for liquid argon were undertaken, followed shortly by simulation of liquid water in 1971 (Rahman and Stilling.Fh, 1971). The first polypeptide (small protein) simulations appeared in 1970s (Chou and Fasman, 1974; Levitt and Warshel, 1975; Mccammon et al., 1977; Gelin and Karplus, 1975). Molecular modeling has become as a proper tool for investigating the atomistic mechanisms that control properties at the nanoscale level, and for relating this information to macroscale material properties. The behavior, function, and failure of materials are directly linked to specific atomistic mechanisms and require molecular level modeling as a mandatory tool for studying biological materials. Unlike a bulk metal (such as gold, for example), information on the specific amino acid sequence that defines protein material allows the researcher to track the resulting effect on structural and system-level behavior. A theoretically-driven understanding of molecular mechanisms for different material systems is crucial for a complete description of the materiome, and thus description of molecular dynamics methods is the central focus of this chapter. However, for the sake of brevity, we focus only to the key theoretical basis of molecular simulation methods, including a limited selection of force fields. In the strict sense, the aim of MD techniques is simply to accurately track the motion of a simulated group of atoms, which represent a significant fraction of a larger system, in order to observe a critical phenomenon of interest, and/or to get an evaluation of the global system properties. Differently to statistical mechanics, where the microscopic properties of individual atoms and molecules are related to the macroscopic bulk properties of a system via thermodynamic metrics, in MD we consider the evolution of a system in the course of time, and the results of molecular dynamics simulations may be used to determine macroscopic properties of the system relying on the ergodic hypothesis, that is statistical ensemble averages are equal to time averages. In this view, it is clearly important that extraction of meaningful data through MD simulations requires significant averaging of the quantities that are sampled (to remove statistical noise), in a similar way to the collective ensemble behavior of statistical mechanics. In other words, like most science, a single result is not enough to confirm a hypothesis, instead truth surfaces over several repeated trials.

The fundamental machinery behind MD is an interaction potential that describe attractive and repulsive forces in between groups of atoms. The interaction potentials (also called force fields) are in general based on a mix of empirical data and first-principles information (based on quantum mechanics calculations, see Fig. 2). Solving each atoms equation of motion according to Newtons Law $F = ma$, enables the calculation of positions $r_i(t)$, velocities $v_i(t)$, and accelerations $a_i(t)$ at each step in time, leading to atom trajectories that can reveal overall dynamics of the system as well as properties such as elasticity, viscosity, or fracture toughness.

Formally, the total energy of the system is written as the sum of kinetic energy (K) and potential energy (U):

$$E = K + U \tag{1}$$

where the kinetic energy is given by

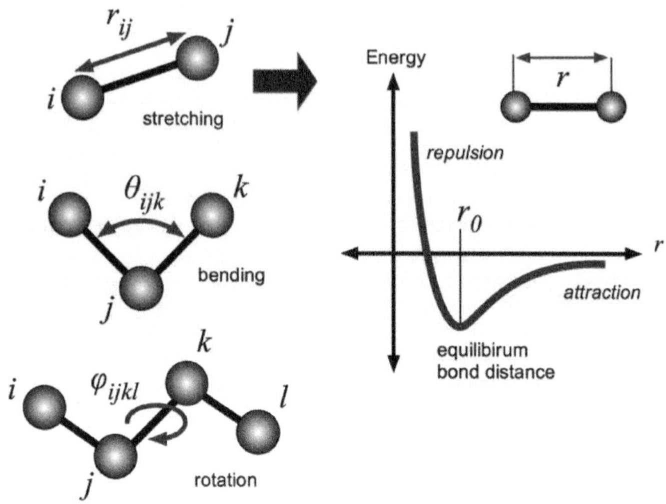

Figure 2. Illustration of the energy decomposition in classical molecular dynamics force fields, along with a representation of a simple potential function between pairs of atoms.

$$K = \frac{1}{2}\sum_{j=1}^{N} m_j \overline{v}_j^2 \tag{2}$$

and the potential energy is an effective function of the atomic coordinates, $\overline{r}_i(t)$, such that:

$$U = \sum_{j=1}^{N} U_j(\overline{r}_j) \tag{3}$$

with a properly defined potential energy surface, $U(\overline{r})$. Here, we emphasize the *effective* potential energy as a function of coordinates (or position), because more sophisticated atomistic force fields take into account not only position, but also environmental effects (such as atomic neighbors and local charge). In general, the potential, U can be formulated as a function of multiple geometric constraints (summed for each atom), including bond stretching, $U(r_{ij})$, bending, $U(\theta_{ijk})$, and rotation or torsion, $U(\phi_{ijkl})$ (See Figure 2). Regardless of form, the forces are obtained from the potential energy surface (sometimes also called force field or potential) as $\overline{F} = -\nabla_{\overline{r}_j} U(\overline{r}_j)$. Simplification of atomic interactions by a closed-form potential significantly increases the computational efficiency of MD simulation. Thus, through Newton's Second Law, molecular dynamics is reduced:

$$m_j \frac{d^2 \overline{r}_j}{dt^2} = -\nabla_{\overline{r}_j} U(\overline{r}_j) \tag{4}$$

For $j = 1 \ldots N$. The numerical problem to be solved is thus a system of coupled second-order nonlinear differential equations that can only be solved numerically given more than two particles (*i.e.*, the classical n-body problem). Typically, MD is based on updating schemes that yield new positions from the old positions, velocities, and the current accelerations of particles, as well as any imposed boundary conditions, forces, of constrains. For instance, in the commonly used Verlet scheme, this can be mathematically formulated as:

$$r_i\left(t_0 + \Delta t\right) = -r_i\left(t_0 - \Delta t\right) + 2r_i\left(t_0\right)\Delta t + a_i\left(t_0\right)\left(\Delta t\right)^2 + O\left(\Delta t^4\right) \tag{5}$$

The forces and accelerations are related by $a_i = f_i/m$. The forces are obtained from the potential energy surface (or force field) as:

$$F = m_j \frac{d^2 \overline{r}_j}{dt^2} = -\nabla_{\overline{r}_j} U(\overline{r}_j) \tag{6}$$

For $j = 1 \ldots N$. This technique - the time integration of particle positions based on applied potentials - cannot only be used for particles that are atoms, it can also be applied for particles that represent groups of atoms, such as in molecules or in coarse-grain models.

Various fast numerical integration schemes are employed to solve the equations of motion and simulate a large ensemble of atoms representing a larger material volume; however, in particular for all-atom simulations, high-frequency vibrations of light atoms (such as hydrogen atoms) requires a time step in the order of femtoseconds (10^{-15} seconds!) for accurate and numerically stable calculations. This limits the application of full-atomistic MD methods to nanometer size systems, at submicrosecond time scales. It is also noted, however, that simulations in excess of hundreds of nanoseconds typically run for weeks or months and large computer clusters with hundreds of processors.

3 Modeling chemical bonds: the force field

A fundamental choice when running molecular simulations is the selection of the potential energy function to describe atomic interactions. The force field is at the very crux of molecular dynamics technique, since it defines the whole material behavior. The potential energy functions are used to determine the forces and accelerations used in a MD updating scheme. Several types and declinations of force fields are available, ranging from those parameterized from detailed quantum calculations to empirically derived formulations, and they are generally implemented within a wide range of simulation programs which are continuously being refined and expanded. Among the most used MD packages, GROMACS (GROningen MAchine for Chemical Simulations)(Van der Spoel et al., 2005) package is commonly used in atomistic molecular dynamics of biomolecules. The NAMD (Not just Another Molecular Dynamics) (Nelson et al., 1996) program is a widespread code that is capable of carrying out simulations using CHARMM (Chemistry at HARvard Macromolecular Mechanics) and other force fields, specifically focused on biomolecules. On the other hand, the LAMMPS (Large-scale Atomic/Molecular Massively Parallel Simulator) molecular dynamics code is a very flexible program that has potentials for soft materials (*e.g.*, biomolecules and polymers), solid-state materials (*e.g.*, metals, semiconductors), as well as for coarse-grained systems (Plimpton, 1995). LAMMPS can be used to model atoms but, more generically, is capable of to work as particle simulator at the atomistic, meso, or continuum scales. In the following we briefly describe some of the basic force field components and concepts, applied in material systems that range from proteins to metals to biomolecules. The methods developed within the framework of conventional force fields are widely applicable, and are the basis for coarse-grain descriptions of molecular behavior. Among the most used force fields specific to protein and biomolecules we find the CHARMM, DREIDING, AMBER and OPLS force fields, widely used throughout the computational biologists and biophysics community. These type of force fields fall into a general class of nonreactive potentials. Simply speaking, a nonreactive potential is not able to consider chemical reactions, as it feature permanent molecular connectivity and chemical state. These force fields are constructed with harmonic and anharmonic terms describing covalent interactions, in addition to long-range terms that describe van der Waals (vdW) interactions, ionic (Coulombic) interactions, as well as hydrogen bonds. Since the bonds between atoms are modeled by harmonic springs or similar approximations, bonds between atoms cannot be broken, and new bonds cannot be created, nor can connectivity configurations be altered. Further, the atomic partial charges are fixed and cannot change. That said, nonreactive force fields provide a good compromise between efficiency and a reasonable description of the behavior of biomolecules. For example, a recent study used the CHARMM force field to study the transition from α-helical to β-sheet secondary structure (α-β transition)a universal deformation mechanism in protein materials such as wool, hair, hoof, and cellular proteins (Qin and Buehler, 2010). Since the α-β transition does not involve any chemical reactions, the nonreactive force field can accurately capture the change in protein conformation. Furthermore, since the transition is due to an externally applied force, MD formulations are required in order to capture the dynamic evolution of the system. An important contributing aspect to the accuracy of such nonreactive force fields is the relatively limited number of parameters to consider*i.e.*, all combinations of atom interactions can be accounted for with rather minimal effort. In the following we discuss the main aspects of the CHARMM force field; however, the basic concepts of the MD technique and force field formulations are common to all MD codes and force fields. The parameters in potential energy function are often determined from quantum chemical simulation models and experimental data in the framework of force field training. As an example, parameters for the CHARMM force field have been accurately improved and revised over the years, taking into consideration for its development a wide variety of input which include: *ab initio* results (*e.g.*, Density Functional Theory), experimental crystal structures and geometries, as well as vibrational spectra (MacKerell et al., 1998). Precisely, the CHARMM potential includes bonding and non-bonding interaction terms to describe forces between particles, where the contributions to bond stretching, bending, and rotation are separately expressed and described by simple mathematical expressions (Fig. 2). For example, both bond stretching and bond angles are described by a simple harmonic spring functions:

In addition to simple harmonic parameters (*e.g.*, k_{bond} representing bond *stiffness*, r_0 bond *length*), several other terms are included to model the chemical properties of proteins and nucleic acids correctly. In the CHARMM model, for example, the mathematical formulation for the empirical energy function that contains terms for both internal and external interactions has the form:

$$
\begin{aligned}
U_{CHARMM} &= U_{bonds} + U_{UB} + U_{angle} + U_{dihed.} + U_{imp.} + U_{vdW} + U_{Coul} \\
&= \sum_{bonds} K_b (b - b_0)^2 \\
&+ \sum_{UB} K_{UB} (S - S_0)^2 \\
&+ \sum_{angles} K_\theta (\theta - \theta_0)^2 \\
&+ \sum_{dihed.} K_\chi \left[1 + \cos(n\chi - \delta)\right] \\
&+ \sum_{imp.} K_{imp} (\phi - \phi_0)^2 \\
&+ \sum_{pairs} \epsilon \left[\left(\frac{\sigma}{r_{ij}}\right)^{12} - \left(\frac{\sigma}{r_{ij}}\right)^{6} \right] \\
&+ \sum_{ions} \frac{q_i q_j}{\epsilon_d r_{ij}}
\end{aligned}
\qquad (7)
$$

where K_b, K_{UB}, K_θ, K_χ, and K_{imp} are the bond, Urey-Bradley, angle, dihedral angle, and improper angle force constants, respectively; b, S, θ, χ, and ϕ are the bond length, Urey-Bradley 1,3-distance, bond angle, dihedral angle, and improper torsion angle, respectively, with the subscript zero representing the equilibrium positions for the individual terms.

The Coulombic (*e.g.*, ionic) and Lennard-Jones 6:12 terms constitute the external or nonbonded (pair) interactions; ϵ reflects the strength of the interaction, or the Lennard-Jones well depth, σ is the distance at the Lennard-Jones minimum, q_i is the partial atomic charge, ϵ_d is the effective dielectric constant, and r_{ij} is the distance between atoms i and j. In the CHARMM force field, there are no explicit terms for H-bonds, since the combination of charge and Lennard-Jones contributions are adequate for appropriately describing hydrogen bonding. In all-atom force fields, solvent (water molecules) is generally treated explicitly and parameters of the force field generally are associated with a specific water model (*e.g.*, TIP3P model for CHARMM (Mackerell, 2004; Ponder and Case, 2003)). The CHARMM force field fits into a class of models with analogous descriptions of the interatomic forces, alongside with other models include the DREIDING force field, the UFF force field (Universal Force Field) (Rappe et al., 1992), or the AMBER model (Wang et al., 2001; Pearlman et al., 1995). As described above, in CHARMM and similar classical force fields, bonded interactions are modeled with harmonic springs or similar formulations, and thus cannot be altered -for example towards a different chemical state, such as from sp2 to sp3, or broken- once defined by the connectivity input obtain from the initial topology of the molecule. In addition, the atomic charges are fixed and cannot change during the course of the molecular simulation. These simplifications drastically increase the simulation speed and do not represent major problem for most simulations, in particular when studying conformational changes of molecules under mild physiological environments. On the other hand, simulations in extreme environments such as mechanical stress (*e.g.*, protein unfolding studies that reach the breaking of covalent bonds) or modeling the properties under the exposure to harsh chemical environments may need the use of reactive force fields which can accurately model changes in the atomic charges as well as the breaking or formation of new covalent bonds and modifications in the bond *orde*r (e.g., single versus double bond). We refer the interested reader to review papers for further information, in particular concerning force field models (Mackerell, 2004; Wang et al., 2001; Scheraga et al., 2007; Deniz et al., 2008; Buehler and Keten, 2010).

4 Towards the micro-scale: coarse-grain modeling

Molecular modeling techniques that rely on atomistic simulations have been excessively (computationally) expensive when long simulation time and/or large-size systems are involved. This led to the advance of so called coarse-grained approaches (Tozzini, 2005). In this framework, by using formulations and computational algorithms similar to full atomistic molecular dynamics, coarse model particles can be simulated using simple bonded and nonbonded interactions that can be easily calculated at each integration time step. The combined usage of simulation procedures with diverse computational cost and precision is denoted to as multiscale modeling, where a orderly link is established between multiple time and length scales. Though we emphasize on the coarse-graining of full atomistic systems, in general multiscale modeling embraces any systematic passing of information from lower levels to larger (coarser) scales. Coarse-grain methods can be generally described as any model formulation that simplifies *many* degrees of freedom to *few* degrees of freedom. The requirement for such approaches arise when the properties of interest have influences (causes or effects) extending over multiple length and time scales which is characteristic of several biological materials. This reduction in degrees of freedom is not novel: simplified systems for investigation can be found in many engineering areas. For example, complex electronic components are engineered based on simplified models of circuits, where the behavior of the key element is defined by general properties such as current, voltage, and resistance. Robust building constructions are evaluated through concepts of beam deflections and beam-column joint rotations, amongst other simplifications. In both examples, more exhaustive system representations are available and can be employed (*e.g.*, influence of temperature and material effects in a transistor, or detailed analysis of stress concentrations in bolts or welds). It is evident that such additions produce a more precise depiction of the modeled system, but also will lead to an increase in the computational cost of the analysis as well as introduce a more sophisticated theoretical framework (which subsequently requires a more detailed set of material and model parameters). However, rarely the use of (simplified) computationally efficient models is justified only by inaccessible time- and length-scales of the more detailed description. Instead, such models are used for analysis *in lieu* of a more complete model because they offer an accurate representation of the system behavior. In the framework of force field training, full atomistic simulations are used to test and acquire the behavior of small molecular components and then to parameterize coarse-grain force fields. These potentials, in turn, are implemented in coarse-grain representations that can facilitate the analysis of large-scale system level phenomena. There is an important inherent assumption, that the coarse-grain theoretical framework provides an accurate depiction of the atomistic behavior, similar to the case where application of elastic beam theory assumes the reliability of a steel girder within a structure. Even in circumstances where full atomistic representations are computationally likely, a coarse-grain description can still be useful for systematic analysis of variable system configurations, which requires a high number of simulations. There is also a further motivation for coarse-graining approach (in addition to the extension of accessible time- and length-scales) and is where the catalyst for coarse-grain potential development is not the extension of atomistic molecular dynamics, but instead to provide an accurate and reliable method for system-level analysis and probe the structure-property and mechanical response for hierarchical systems. Furthermore, mesoscopic models provide simple and efficient modeling techniques for experimentalists, permitting a more direct comparison between simulation and a vast variety of experimental methods, without requiring specialized molecular dynamics tools such as high-performance computer clusters with complex hardware and software. A developed coarse-grain model can only reflect the behavior included in the governing potentials and associated parameters, and therefore, the origin of these parameters naturally determines the accuracy and effectiveness of the coarse-grain model. A complete theoretical foundation of the *materiome* requires synergistic multiscale transitions from atomistic to mesoscopic to macroscopic descriptions. Hierarchical handshaking at each scale is critical to predict structure-process-property relationships, to provide key mechanistic understanding, and to enable predictive modeling and material optimization to guide design efforts. As proposed, a *finer-trains-coarser approach* is not limited to link atomistic to mesoscopic scales, but can refer to parameterization transcending any scales, such as atomistic to continuum levels, or beam-columns to buildings. Such a multiscale modeling paradigm establishes a fundamental link between constituent behavior and system-level function, providing a consistent theoretical method to develop coarse-grain models for systems of various scales, constituent materials, and intended applications (Fig. 3).

More computationally expensive (and sophisticated) techniques (*fine*), such as Density Functional Theory (DFT), are simplified in the framework (parameters and functions) in classical atomistic potentials (*coarse*). In this view, the sub-atomic structures (bosons, fermions, etc.) can be seen as the material building blocks, while bonded atoms can be thought of as the structural system. From this standpoint, atomistic force fields can be looked as a systematic and efficient coarse-graining of quantum mechanics. Indeed, the direct application of quantum mechanical methods to large macromolecular systems (such as biomolecules) is not yet feasible due to the large number of atoms involved. As a logical extension, the output of atomistic simulations can be implemented in the parameterization of mesoscopic coarse-grain potentials. Each transition involves an increase in accessible system size, length scale, and time-scales, as well as a simplification of the governing theoretical models. As a consequence, details of interactions are lost, but the desired fundamental behavior is preserved. It is noted that the desired fundamental behavior depends on the particular application, and can include molecular interactions, mechanical behavior, thermal or electrical properties as well as equilibrium configurations. For example, an atomistic depiction of carbon bonding can include terms for atom separation, angle, and bond order (such as CHARMM or AMBER type force fields), but do not account for electron structure, band gaps, or transition states found in quantum mechanical methods. Nevertheless, the atomistic model maintains the correct behavior of the individual carbon atoms, and neglecting the effects of electrons is considered a necessary simplification. Similarly, the coarse-graining of a carbon nanotube integrates the effects of multiple carbon bonds into a single potential. The exact distribution of carbon interactions is lost, but the behavior of the molecular structure is preserved. The coarse-graining process discussed hereafter focus on constituent materials that can be modeled by full atomistic methods using classical molecular

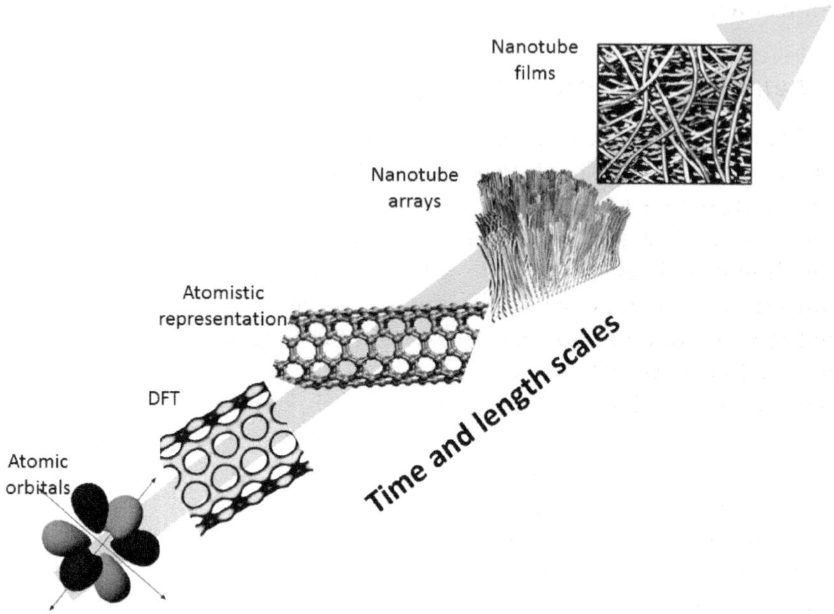

Figure 3. Jumping scales: quantum to atomistic, atomistic to mesoscopic. The rationale for coarse-graining approaches can be justified through analogous comparison between the parameterization of atomistic force fields by detailed quantum mechanical methods to the development of coarse-grain potentials by the results of full atomistic simulations. Atomic orbitals (wave functions) leading to the electron charge isosurface of a carbon nanotube via Density Functional Theory (DFT), wherein the behavior of carbon atoms can be determined. Full atomistic model of identical carbon nanotube using carbon-carbon interactions allowing the simulation of larger molecular structures. Coarse-grain carbon nanotube systems, using potentials developed from atomistic results. Systems consisting of hundreds of nanotubes can be represented, reaching scales on the order of micrometers (and implicitly representing billions of atoms). Each transition involves an increase is accessible system size, length scale, and time scales, as well as a simplification of the governing theoretical models. Adapted from (Cranford and Buehler, 2012).

dynamics, while the system structure (and relevant behaviors) requires a fully informed coarse-grain representation. The atomistic model provides the underlying physics to the coarse-grain potentials, allowing investigation at the desired length scale. The advance of coarse-grain models allows the modeling of events on physically relevant time and size scales, leading to a range of possible advances in nanoscale design and engineering in a completely integrated bottom-up methodology. A coarse-graining approach is intended to develop tools to explore material properties and underlying mechanical behavior typically required for material design that systematically integrates characteristic chemical responses. However, it is stressed that the purpose is not to avoid the need for full atomistic simulations. As a matter of fact, coarse-graining *requires* accurate full atomistic representations to acquire the required potential parameters. In conclusion, the finer-trains-coarser procedure described here is an attempt to reunite first principles formulations with hierarchical multiscale methods by using atomistic theory with molecular dynamics simulations *in lieu* of empirical observations, in a integrated and systematic framework.

5 Modeling collagen tissues - Collagen hierarchical structure

The word collagen originates from the Greek word $\kappa o\lambda\lambda\alpha$ (*kolla*, meaning glue) due to its ancient use as a glue source - obtained though the boiling of animal skin and collagen-rich tissues - from Egyptians about 4,000 years ago (Chamberlain et al., 2011). In a broader sense, collagen could be seen as the glue of our body, holding it together by providing elasticity and strength to several tissues where a mechanical functions is needed, such as skin, cartilage, tendons and bones (Fratzl and Weinkamer, 2007; Fratzl, 2008). From a structural point of view, collagen-rich tissues are built with the collagen fibril as fundamental building block. These fibrils have a diameter in the range of 30 to 500 nm, a length up to the millimeter range and are assembled in complex hierarchical assemblies, whose structure depend on the particular tissue (Fig. 4).

At the lowest hierarchical level there is the collagen molecule (also called *tropocollagen*), that is a rod-like molecule with a length of about 300 nm and a diameter of about 1.5 nm, made of three separate chains folded in the characteristic triple helical configuration (Ramachandran, 1955; Rich and Crick, 1955). According to the Hodge-Petruska model (Petruska and Hodge, 1964)a structural model of collagen fibrils proposed in 1964molecules in a fibril are deposited side by side and parallel but staggered with respect to each other, where the molecular axes are also parallel to the fibril direction.

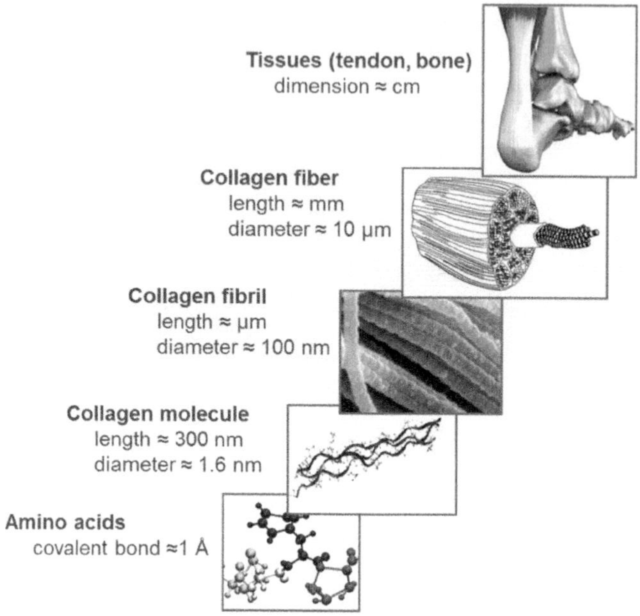

Figure 4. Hierarchical structure of collagen protein materials. Each collagen molecule is made of three peptide chains that form the ≈300 nm long triple helical collagen molecule. Collections of collagen molecules aggregate both in lateral and longitudinal directions to form fibrils. Fibrils in cornea are normally thin (≈30 nm) and uniform in diameter, while tissues such as tendon contain a wide-ranging distribution of diameters (100-500 nm). Fibrils include tiny hydroxyapatite crystals in bone tissue, which provide stiffness and compressive load resistance. In tendons and ligaments multiple fibrils make up collagen fibre, formed with the aid of proteoglycans. Adapted from (Gautieri et al., 2011).

This structure creates an observable periodicity known as the D-period whose length is D=67 nm, in which two region can be observed, the gap and overlap regions, emerging due to the fact that the length collagen molecule is not a multiple of the D-period itself (Fig. 5). A particularly important feature for the mechanical properties of fibrils is the presence of enzyme-induced covalent cross-linking between molecules, which develops only with maturation of the tissues (Eyre et al., 2008).

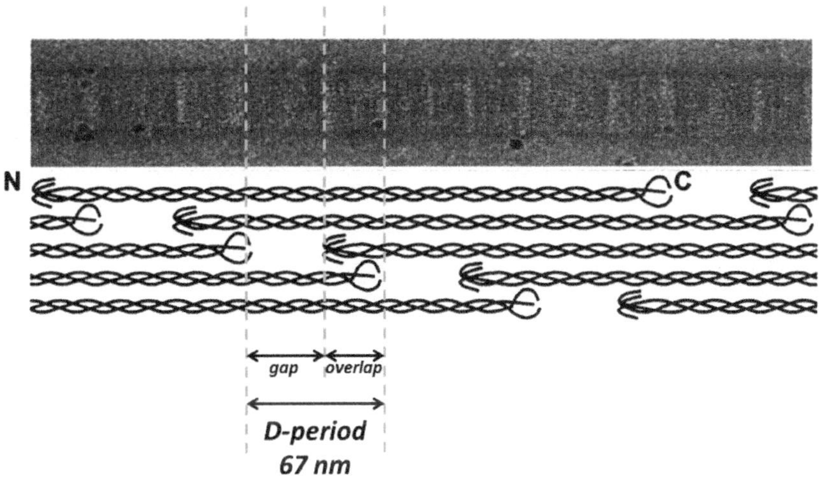

Figure 5. D-period of collagen fibrils. Electron microscope of a collagen fibril (top panel) and schematic representation of the axial arrangement of molecules in the D periodic fibril (bottom panel). Adapted from (Gautieri et al., 2011).

At the next level of the hierarchy, the organization of the collagen structure varies greatly depending on the tissue. In tendons and ligaments, multiple fibrils make up the collagen fiber, which are formed with the aid of cross-linking macromolecules such as proteoglycans. In bone, the organic collagen protein matrix alone is not sufficient alone to provide the stiffness and resistance to compression required for this tissue (which has to carry considerable loads) and additional stiffening is provided by the inclusion of mineral hydroxyapatite crystals into collagen fibrils and particularly in the gap region (Fratzl, 2008; Currey, 2002; Weiner and Wagner, 1998). Skin and artery walls need to be flexible but arteries also need to resist blood pressure. This implies yet a different arrangement of collagen fibrils in combination with elastomeric molecules, such as elastin.

Due to the biological importance of collagen-based tissues there have been a wealth of investigation concerning the relationship between structure and properties in these complex materials, starting from early works (dated in the 50s) on of the structure of collagen molecules (Ramachandran, 1955). However, there are still open questions (*e.g.*, concerning the mechanism of load transfer between the hierarchical scale of bone and tendons, or the onset of collagen-related diseases). A very promising technique to complement experiments in addressing these questions is represented by multi-scale modeling. In the following of this Chapter, successful application of atomistic and coarse-grain modeling techniques for the study of collagen materials are presented.

6 Visco-elastic properties of collagen molecules

In vertebrates, locomotion and movements are achieved through the generation of muscular forces that are then transmitted to joints. The force transmission (mechanotransduction), which involves the storage, release and dissipation of energy, is provided by connective tissues such as ligaments and tendons (Alexander, 1983, 1984). Thus, in these tissues viscoelasticity is a fundamentally important feature, as both viscous (time-dependent) and elastic (time independent) contributions determine how ligaments and tendons accomplish their function as mechanotransducers. The mechanical behavior of these connective tissues is directly related to their complex hierarchical structure and to their specific macromolecular components (Silver et al., 2001). The lowest hierarchical scale of several load-bearing tissues is represented by collagen type I molecules (Kadler et al., 2007). At the macro-scale (*i.e.*, whole tissues and collagen fibers), the mechanical investigation of collagen-based tissues has been performed for several decades. As a consequence, it is well known that collagen-rich tissue present viscoelastic behavior, as established from a larger number of creep and relaxation tests that have been performed on both tendons and ligaments (Wang and Ker, 1995; Rigby et al., 1959; Sasaki et al., 1999). Based on experimental data, and starting from the pioneering work of Fung (Fung, 1967), several constitutive and structural models have been developed to model the viscoelastic behavior of collagen at the tissues scale (Haut and Little, 1972; Puxkandl et al., 2002; Pena et al., 2008; Screen et al., 2011). On the other hand, fewer studies focused on the viscoelastic properties at smaller scales. At the fibril level, earlier investigations have been performed through small angle X-ray scattering (Sasaki and Odajima, 1996) to assess the fibril deformation inside tissue specimens. However, in this setup the load is not directly applied to single fibril and hence the response is averaged over several fibrils. Direct experimental studies of single fibrils are relatively recent and made possible due to the use of atomic force microscopy or micro-mechanical systems which allowed the study of the elastic (van der Rijt et al., 2006; Svensson et al., 2010b; Shen et al., 2008; Eppell et al., 2006; Shen et al., 2010) and viscoelastic properties of collagen fibrils (Shen et al., 2011; Svensson et al., 2010a). Ultimately, at the molecular level several studies have been performed using both experimental and computational procedures; however most of these investigations focused solely on the elastic properties (Sasaki and Odajima, 1996; Sun et al., 2002; Buehler, 2006a; Gautieri et al., 2008, 2009, 2010). For this reasons, despite the abundant investigation of the viscoelastic behavior at the macro-scale and the very recent works on collagen fibril viscoelasticity, relatively little is known about the molecular and fibrillar origin of the time-dependent properties of collagenous tissues. It has been speculated that the elastic behavior of these tissues is due to the stretching of cross-linked collagen molecules and thus fibrils, whereas the energy dissipation (the viscous behavior) is thought to involve sliding of molecules and fibrils by each other during the tissue deformation (Silver et al., 2001). However, there is still no clear understanding of the mechanisms behind the viscoelastic behavior of collagenous tissue, and on the role of each hierarchical scale in determining the overall mechanical properties. An important outstanding question is whether or not molecular-level load relaxation effects are important for the viscoelastic properties measured at larger scales.

An approach to investigate the molecular origin of collagen viscoelasticity, is the use of *in silico* tests to study collagen molecules from the level of amino acids upwards (Gautieri et al., 2012). In this way it is possible to obtain quantitative data on the time-dependent mechanical behavior of single collagen molecules in order to assess molecular-level viscoelastic properties, and to scale up properties from there to larger tissue levels. In this approach, Steered Molecular Dynamics simulations is used to perform *in silico* creep tests (*i.e.*, constant load tests) of collagen molecules, fitting the deformation response over time with a Kelvin-Voigt (KV) model. The KV model is a simple constitutive model consisting of a spring and a dashpot in parallel (see Fig. 6), where this approach allows us to compute the Youngs modulus, the viscosity and the characteristic relaxation time of a single collagen molecule. In order to perform constant force SMD simulations, the

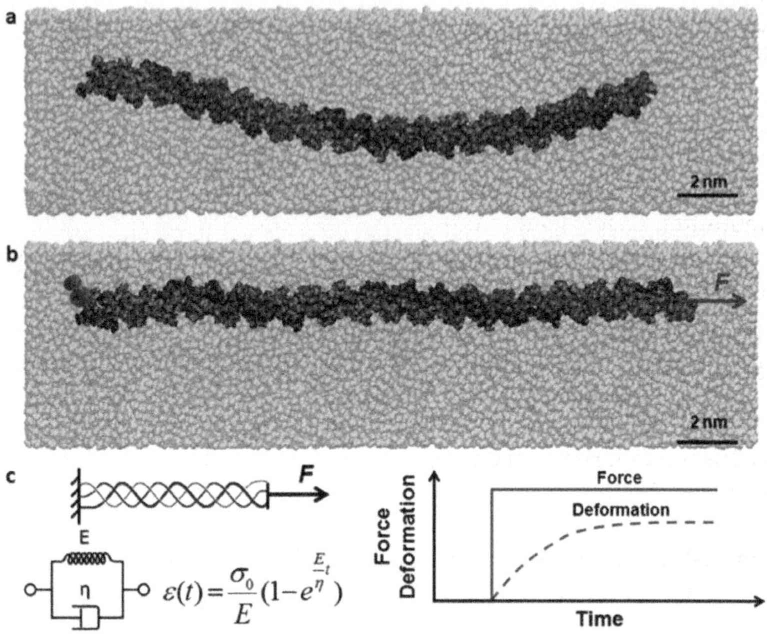

Figure 6. Snapshots of the collagen peptide in water box. Panel a shows the conformation of the full atomistic model of a collagen peptide solvated in water box and equilibrated for 30 ns. After equilibration the molecule is subjected to virtual creep tests: one end of the collagen peptide is held fixed, whereas the other end is pulled with constant force (from 300 pN to 3,000 pN) until end-to-end distance reaches equilibrium (Panel b). Panel c shows a schematic of the creep test; a constant force is applied instantaneously to the molecule and its response (deformation over time) is monitored. The mechanical response of collagen molecule is modeled using a KV model, from which molecular Youngs moduls (E) and viscosity (η) are calculated. Adapted from (Gautieri et al., 2012).

three N-terminal C_α atoms are kept fixed whereas the three C-terminal C_α atoms are subject to an instantaneous constant force. This setup mimics a single molecule creep test, in which an instantaneous load is applied and the strain response in time is observed. The creep test simulations are run until asymptotic deformation and then strain-time curves obtained for collagen molecules during SMD simulations are fitted with the Kelvin-Voigt equation in order to estimate E (Youngs modulus) and η (viscosity). The instantaneous stress σ_0 is calculated dividing the applied force by the cross-sectional area of collagen, assuming a round cross-section and a radius of 0.55 nm. The value of the radius is determined from dry collagenous tissue in which the lateral distance between molecules is found to be 1.1 nm (Fratzl et al., 1993).

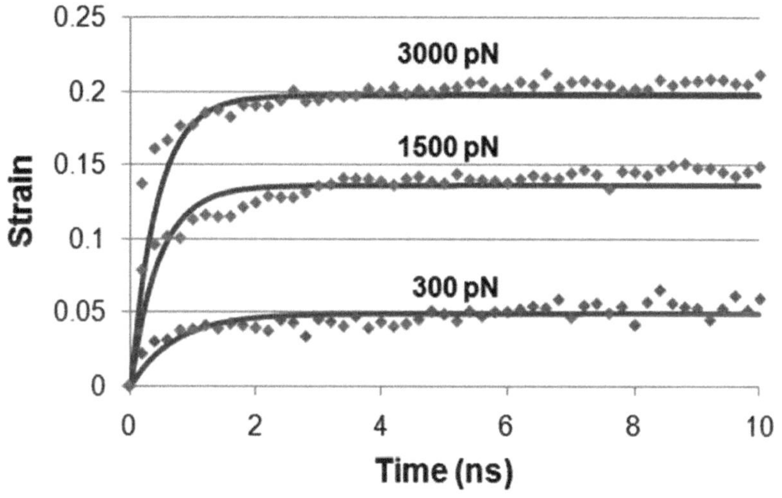

Figure 7. Single molecule creep test. Mechanical response of solvated collagen molecule to creep tests for three cases with increasing value of external force. Dots represent the experimental data, whereas curves represent the fitted curves using a KelvinVoigt model. Adapted from (Gautieri et al., 2012).

When subjected to creep tests, the engineering strain of the molecule shows an exponential increase until asymptotic value is reached (see Fig. 7). This is the behavior typical of simple viscoelastic materials that can be characterized by a Youngs modulus E, responsible for the elastic response, and a viscosity , responsible for the viscous behavior. These results show that collagen molecule presents a non-linear elastic behavior, since the Youngs modulus depends on the external load. In particular, for the loading conditions considered here E ranges from 6 to 16 GPa, a finding that matches several experimental (Sasaki and Odajima, 1996; Sun et al., 2002) and modeling studies (Gautieri et al., 2009; Srinivasan et al., 2009). The collagen molecule viscosity has an average value of 3.84.0.38 Pa·s and it is not significantly affected by the external load. Based on the elastic modulus and the viscosity, the characteristic relaxation time (defined, for a KV model, as η/E) is calculated, obtaining values in the range 0.24-0.64 ns and 0.13-0.27 ns for solvated and dry molecule, respectively. Although the elastic behavior has been previously investigated, no experimental data is available for the viscous component of a single collagen molecule. This could be attributed to the very fast relaxation time (on the order of nanoseconds) of single molecules, which hinders its assessment using available experimental techniques. However, a recent work (Nakajima and Nishi, 2006) demonstrated that Atomic Force Microscopy (AFM) can be successfully used to assess the viscoelastic properties of a single polypeptide chain, in this case polystyrene. Within this technique, referred to as nanofishing by the authors of that study, an AFM cantilever is mechanically oscillated at its resonant frequency during the stretching process. This enables the estimation of stiffness and viscosity of a single polymer chain with the use of a phenomenological model (the authors used a KV model). The polystyrene viscosity per unit length reported by the authors is $2.6 \cdot 10^{-9}$ Kg/s, which agrees well with the findings for collagen molecules, *i.e.*, $40 \cdot 10^{-9}$ Kg/s (considering a peptide length of 20 nm). The slightly higher viscosity could be attributed to the fact that the polystyrene chain is a single strand, while the collagen peptide is in a triple helical configuration. It is of great interest to discuss whether the two elements of the KV model, *i.e.* the purely elastic spring and the purely viscous dashpot, have an actual physical meaning. A likely explanation would be that the elastic spring corresponds to the protein backbone, while the damping effect could be attributed to the interchain H-bonds. The backbone deformation include dihedral, angle and bond deformation, which are terms expressed by harmonic (or similar) functions in the molecular dynamics force field, and thus result in an elastic response to stretching. On the other hand, the viscous behavior may be due to the breaking and reforming of H-bonds, in particular H-bonds between the three collagen chains. Statistical mechanics based theories, and in particular Bells model (Bell, 1980), has been applied to study the breaking of H-bonds (see, *e.g.* reference (Buehler and Wong, 2007) and Fig. 8). It has been shown that the velocity at which a H-bond breaks is a function of the external force:

$$v = \frac{x_b}{\tau_0} e^{(-[(E_b - F x_b cos)/(k_b T)])} \tag{8}$$

where v is the velocity at which the H-bond breaks, x_b is the distance between the equilibrium state and the transition state, τ_0 is the reciprocal of the bond natural frequency, E_b is the bond energy, F is the applied force, θ is the angle between the direction of the reaction pathway of bond breaking and the direction of applied load F, k_b is the Boltzmann constant and T is the temperature. The equation above shows that the mechanical response of the H-bonds is time-dependent, matching the behavior of the dashpot in the KV model very well (for which the stress is a function of the loading rate). Therefore the time-dependent rupture of H-bonds are likely responsible for the dissipative behavior observed in collagen molecule mechanics.

7 Atomistic modeling of collagen fibrils

In order to complement experimental approaches, molecular modeling provides a powerful approach to describe the molecular mechanics of collagen. Earlier studies were based on short collagen-like peptides obtained from x-ray crystallography (Beck et al., 2000; Bella et al., 1994; Redaelli et al., 2003). These early molecular simulation studies used these short collagen molecules (Buehler, 2006b; Gautieri et al., 2009; Srinivasan et al., 2009; Lorenzo and Caffarena, 2005; Bhowmik et al., 2007; Gautieri et al., 2009; Bhowmik et al., 2009) that were typically limited to less than 10 nm length or more than a factor of 30 smaller than actual molecules found in collagen tissues. The resulting elastic modulus of these short collagen peptides was found to be on the order of 4 GPa, and thus much greater than the typical Youngs moduli measured for macro-scale collagen tissues but in agreement with single molecule studies (Harley et al., 1977). In order to determine how collagen-based structures confer mechanical properties to tissues like skin, tendon and bone, and to identify how cells interact with the ECM, the understanding of the mechanics at different hierarchical levels and their interplay from a biochemical and molecular level upwards is essential. Earlier work has demonstrated that mechanical strain is distributed over distinct hierarchical levels (molecules, fibrils, fibres) (Gupta et al., 2005) and that collagen tissue stretching involves concurrent deformation mechanisms. The significance of defining the material properties of collagenous tissues from

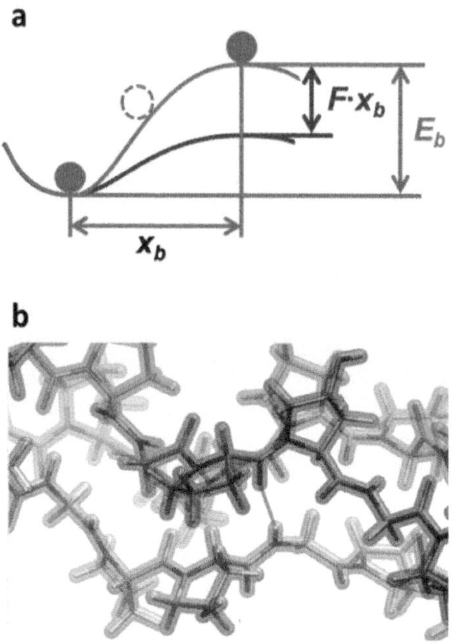

Figure 8. Statistical theory to predict bond rupture mechanics. The graph depicts the energy as a function of a deformation variable (a), along a particular pathway that leads to bond rupture, where E_b is the energy barrier corresponding to the transition state. Panel b shows an interchain H-bond in collagen. Adapted from (Gautieri et al., 2012).

the biochemistry level upwards is evident when considering the effect of mutations in collagen, which can result in incorrectly folded collagen protein that cause a variety of severe and sometimes deadly pathologies, such as Ehlers-Danlos syndrome or osteogenesis imperfecta (brittle bone disease) (Rauch and Glorieux, 2004). Early collagen microfibril and fibril models represent the supramolecular arrangement in collagenous tissues in a simplified way, using a two-dimensional lattice of mesoscopic beads (Buehler and Wong, 2007) or extremely short collagen peptides (Buehler, 2006a; Gautieri et al., 2009; Srinivasan et al., 2009; Lorenzo and Caffarena, 2005; Bhowmik et al., 2007; Gautieri et al., 2009; Bhowmik et al., 2009). These models do not account for the biochemical details and are much smaller than the typical length-scales of collagen molecules found in collagen microfibrils and fibrils. The reason for these approximations was the lack of crystallographic details of the collagen molecule, which have been obtained only for short collagen-like peptides with lengths below 10 nm, until the work of Orgel and coworkers, that obtained the *in situ* structure of full length collagen type I molecule (Orgel et al., 2006) (Protein Data Bank identification code 3HR2) (see Fig. 9). This structure, obtained by employing conventional crystallographic techniques in x-ray fiber diffraction experiments, resolved for the first time the specific three-dimensional arrangement of collagen molecules in naturally occurring fibrils, including the N- and C-telopeptides.

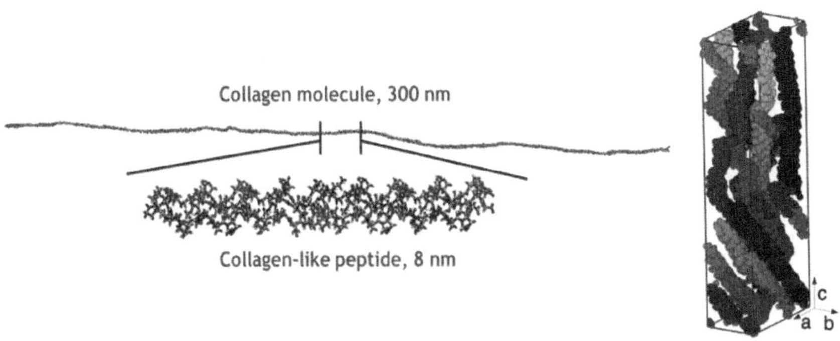

Figure 9. Comparison of collagen-like peptides with the actual collagen molecule (left) and unit cell of the low-resolution X-ray crystallography of the collagen microfibil described in (Orgel et al., 2006).

Starting from the structure reported in (Orgel et al., 2006), which includes only backbone alpha carbons and the primary sequence of *rattus norvegicus*, Gautieri *et al.* (Gautieri et al., 2011) used homology modeling to obtain a full-atom structure with the human collagen sequence (Fig. 10). The molecular packing topology obtained by the periodic repetition of the unit cell lead to quasi-hexagonally packed collagen molecules which interdigitates with neighboring molecules to form a supertwisted right-handed microfibril and to the well-known D-banding periodicity seen in AFM images of collagen fibrils. The fibril model is solvated using the solvate plug in of GROMACS by adding SPC water molecules. Since the molecule at physiological pH includes a net charge (positive net charge +34), counterions (Cl-) are added in order to keep the system neutral. The fibril model is equilibrated through several ns molecular dynamics simulations at a temperature of 310 K (37°C), and with 1 bar pressure to ensures structural convergence. In order to assess the mechanical properties of the atomistic models, molecular dynamics simulations with increasing constant mechanical stress in tension along the fibril axis has been performed, while maintaining the pressure on the other axes constant at 1 bar. The applied stresses are in the range from 0 to 200 MPa, applied during 20 ns molecular dynamics simulation for each load applied.

The equilibrated collagen microfibril (Fig. 11) has a density of 1.19 g/cm^3, a value that is halfway between the density of water and the density of dehydrated collagen, which has been estimated at 1.34 g/cm^3. A Ramachandran analysis of the solvated system shows that the collagen microfibril lies in a region of the diagram (Psi $\approx$150, Phi $\approx$-75) that is characteristic of the polyproline II chain and thus of collagen-like peptides, in good agreement with earlier experimental structural studies (Brodsky and Ramshaw, 1997). When subjected to mechanical load, collagen microfibrils feature two distinct deformation regimes. In the small-strain regime (<10%) the predicted Youngs modulus is $\approx$300 MPa, while in the large-strain regime (>10%) the microfibril shows a severely increased tangent stiffness, with a Youngs modulus of $\approx$1.2 GPa (Fig. 2.13). The results of nanomechanical testing of collagen microfibril are in agreement with experimental results obtained for the small strain regime based on several techniques such as x-ray diffraction (Sasaki and Odajima, 1996), Atomic Force Microscopy (AFM) (van der Rijt et al., 2006) and the use of micro-electro-mechanical systems (MEMS) (Shen et al., 2008; Eppell et al., 2006). For the larger-strain regime there exists less experimental information and available data tend to be more scattered. For example, Shen *et al.* (Shen et al., 2008) showed a relatively large variability of collagen fibril behaviors at large deformation, which suggested either strain-hardening or strain-softening depending on the fibril investigated. A direct comparison of the mechanical properties of single collagen molecules versus that of collagen microfibrils suggests that the mechanical properties are strongly scale dependent (Fig 2.13).

The atomistic model enables to observe atomistic and molecular deformation mechanisms not directly accessible to experimental techniques, and thereby to explain the molecular origin of mechanical properties at different hierarchical levels and levels of strain. It is observed that for small defor-

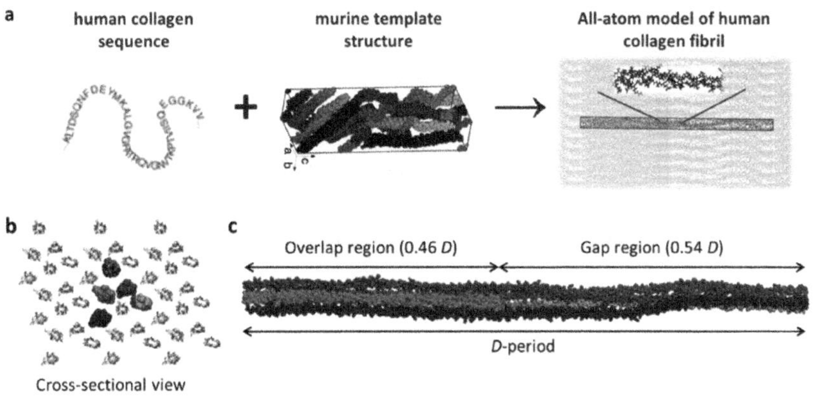

Figure 10. Schematic of the homology modeling process, in which a target sequence is applied to a template structure. The use of the murine template structure is possible thanks to the high sequence homology between human and murine collagen type I (a). Panel (b) shows the quasi-hexagonal cross-section of the microfibril model and panel (c) the D-period divided into gap and overlap regions. Adapted from (Gautieri et al., 2011).

mation the collagen molecules end-to-end distance increases linearly until the microfibril strain reaches 10% (corresponding to ≈50 MPa stress), the strain at which the microfibril stiffness increases drastically. Beyond this point the molecular end-to-end still continues to increase but the slope of curve is significantly lower. This can be explained due to the fact that below 10% strain the collagen molecule is straightened within the microfibril and thereby loses its kinked arrangement, while beyond the 10% strain mark the molecule itself is being stretched, resulting in a larger mechanical resistance to deformation. The increase in the gap to overlap ratio in the small-strain regime (Fig. 11) suggests that the initial straightening is concentrated in the gap regions where the molecular packing density is lower and molecules are less organized, with more kinks. This model thereby confirms sugges-

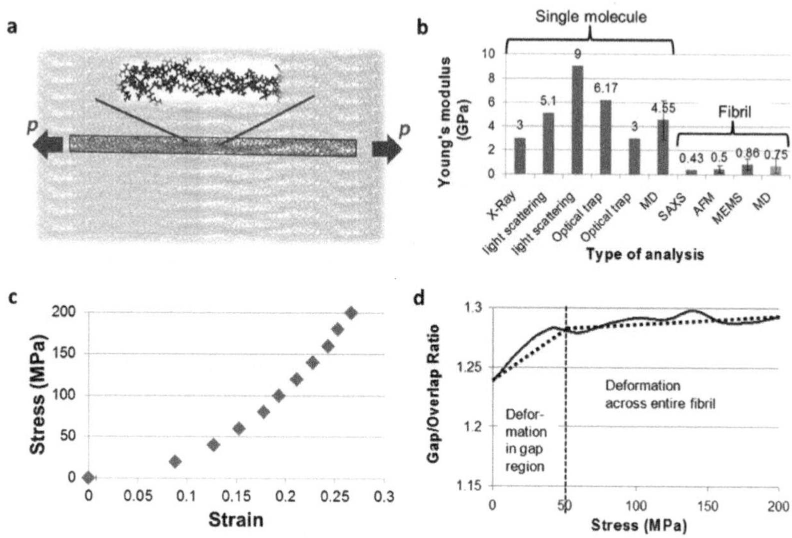

Figure 11. Schematic of the *in silico* mechanical loading (a) and comparison of modeling results with experimental results at fibril and molecular scale (b). Stress-strain relationship for collagen type I microfibril (c) and analysis of the gap/overlap ratio, showing that for small deformation the gap region is deforming more (d). Adapted from (Gautieri et al., 2012).

tions made by Fratzl *et al.* (Fratzl et al., 1998). Once the entire capacity to molecular straightening is exhausted and all collagen molecules assume a straight configuration oriented in the direction of the pulling axis, collagen molecules themselves undergo stretching, leading to significantly increased tangent microfibril stiffness, at strains in excess of 10%. The combination of these two mechanisms, molecular straightening and molecular stretching, effectively lead to an increase of the *D*-period, in agreement with experimental results (Gupta et al., 2005). Other mechanisms, such as molecular sliding may take place at larger strains in excess of 30% as described in earlier studies of the deformation mechanisms (Fratzl et al., 1998; Sasaki and Odajima, 1996).

8 Coarse-grain modeling of collagen

As seen in the previous paragraph, the study of collagen fibrils and fibers with full atomistic simulations is currently challenging, which represents an important limitation since this scale in the structural hierarchy of collagen is most relevant for physiological function. Other limitations exist with respect to the accessible time-scale, where most full-atomistic simulations are limited by a few hundred nanoseconds. However, many relevant materials phenomena such as tissue deformation and failure (*e.g.* under disease conditions) emerge at much longer time-scales. A promising strategy to overcome these limitations is to decrease the number of degrees of freedom by grouping atoms into pseudo-atoms (or particles) referred to as beads. This represents the basis of the so-called coarse-grained approach, where starting at the nanoscale, it is possible to derive parameters for higher hierarchical levels, up to the macroscale by systematically feeding information from smaller, more accurate to larger, more coarse levels. The development of a coarse-grained model provides a powerful path towards reliable modeling of full-length collagen molecules and its higher level hierarchical structures. Specifically, coarse-grained models allow the study of more complex systems, up to micrometer dimension and millisecond duration. Earlier work of coarse-graining collagen molecules grouped hundreds of atoms into particles (Buehler, 2006a; Gautieri et al., 2008). This level of coarse-graining, however, is at a relatively coarse level where information about biochemical features (*e.g.* amino acid sequence) cannot be represented directly. However, the incorporation of biochemical features is crucial in a computational materiomics approach where material properties are elucidated at multiple scales. Several other coarse-grained models suitable for proteins have been developed and successfully applied (Tozzini, 2005). Particularly, the MARTINI coarse-grained model (Marrink et al., 2007), developed by Marrink and coworkers, was initially applied to membrane lipids, later extended to proteins (Monticelli et al., 2008) and to carbohydrates (Lopez et al., 2009). It has been successfully applied to gain insights about different biological molecules such as membrane proteins, ion channels and liposomes (Sansom et al., 2008). The MARTINI model provides a suitable level of coarse-graining as it retains information about the chemistry specific to the amino acid sequence (as side chains are modeled depending on the type of the amino acids). All amino acids in the MARTINI force field are modeled with a number of beads that varies depending on the steric volume of each amino acid (Monticelli et al., 2008). The general mapping rule is that four heavy atoms (that is, non-hydrogen atoms) are grouped together into one bead. Further, one bead describes the backbone, while others are added to represent the side chain. The number of beads used to model a specific residue therefore varies depending on the dimensions of the side chain of the amino acid. Small amino acids such as glycine or alanine are described by just one bead, while larger amino acids, like phenylalanine, tyrosine and tryptophan are modeled with up to five beads. The model also takes in account the polarity of every bead, described by a letter (P: polar; C: apolar; N: non-polar; Q: charged) and a number (from 1: low polarity to 5: high polarity). Further, a letter is used to characterize a residues hydrogen bonding capability (d: donor; a: acceptor; da: donor and acceptor; 0: none). These bead types, which correspond to atom types in the atomistic modeling framework, are used to describe the non-bonded interactions between beads. For each pair of bead types, a set of parameters for electrostatic and van der Waals potentials is defined. The best choice of particle types for amino acids was obtained by the authors of the original MARTINI force field for proteins by comparing simulation results and experimental measurements of water/oil partitioning coefficients of the amino acid side-chain analogues. However, although the MARTINI force field is suitable for a wide number of applications and its validity has been proven through several simulations and comparisons both with atomistic and with experimental data, it cannot be applied to model collagen molecules in the original formulation. This is because of two reasons. First, the existing formulations of MARTINI lack parameters for hydroxyproline (a non-standard amino acid, found solely in collagen and formed through post-translational modification of proline). Second, it lacks parameters to describe the triple helical configuration of a collagen molecule. Thus, recent efforts have been addressed to overcome these limitations and addresses two major points. In the MARTINI model, proline is modeled through the use of two beads, one for the backbone (bead type C5) and one for the side-chain (bead type AC2) as shown in Fig. 12. Hydroxyproline derives directly from proline via the addition of a hydroxyl group on its side-chain. Therefore, hydroxyproline is also modeled using two beads in the extended MARTINI model. For the backbone bead is represented with the same bead type as for proline. The side chain, however, due to the presence of the hydroxyl group, shows a higher polarity level than proline. This aspect was already demonstrated in the work of Black and Mould, who calculated an index that takes into account the hydrophobicity of all amino acid side-chains, including hydroxyproline. Matching the bead types assigned by Marrink and the hydrophobicity values derived from Black and Mould, the hydroxyproline side-chain bead polarity, in the MARTINI notation, is determined. This parameter ranges from 0 to 1, where 1 is the most hydrophobic amino acid (phenylalanine), while 0 is the most hydrophilic one (arginine). The value reported for hydroxyproline is 0.527, which is found between threonine (0.450) and cysteine (0.680) values. In the MARTINI force field, threonine side chain is modeled as a P1 bead, while cysteine as a C5 bead. Considering this result, hydroxyproline is assigned a P1 value for the hydroxyproline side-chain, since its hydrophobicity value is closer to the value for threonine.

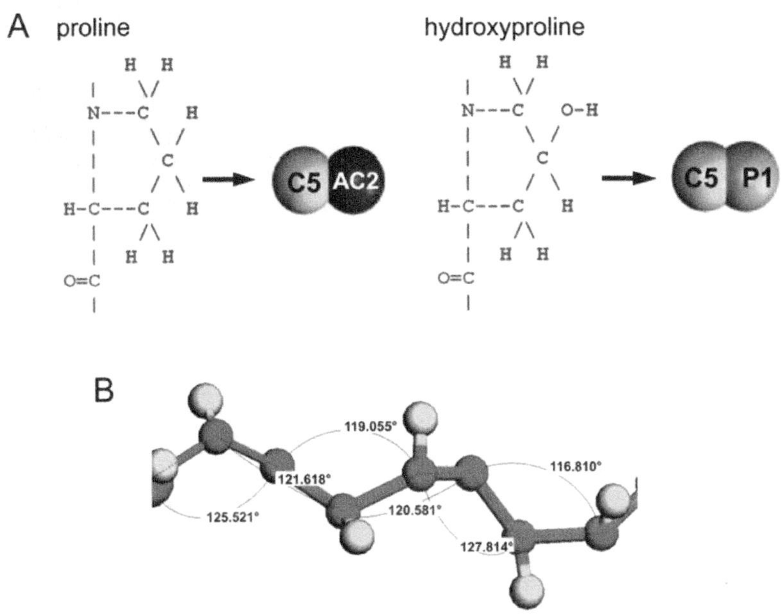

Figure 12. Parameterization of the force field of the MARTINI force field including the hydroxyproline residue (Panel A) and geometrical features (angles) of the coarse-grained collagen triple helix (Panel B). Hydroxyproline directly derives from proline adding a hydroxil group to its side chain, giving rise to an increased polarity level. Atomistic structures of both proline and hydroxyproline are shown on the left side of Panel A, while the bead representation using MARTINI model notation is shown on the right side of Panel A. Panel B shows the analysis of the angle between backbone beads of collagen-like peptides. The statistical analysis is performed on five collagen PDB entry (2DRT, 1K6F, 1QSU, 1CAG, 1V6Q); these collagen-like peptides are coarse-grained and the bond lengths, angles and dihedrals between backbone beads are analyzed. Adapted from (Gautieri et al., 2010).

The MARTINI model also takes into account the secondary structure of a protein, such as alpha-helix, beta-sheet, extended turns, or bends. The characteristic triple helical structure of collagen molecules cannot be described by any of the secondary structure parameters included in original MARTINI model. Thus, the bond distances, angles and dihedral parameters specific to the collagen triple helix have been assessed. The MARTINI bond potential forms are maintained, which are described as follows:

$$V_b = K_b(d_{ij} - d_b)^2 \tag{9}$$

$$V_a = K_\phi(\phi_{ijk} - \phi_a)^2 \tag{10}$$

$$V_d = K_d[1 + \cos(n\chi_{ijkl} - \chi_d)] \tag{11}$$

where V_b, V_a and V_d are the potential energy terms of bond stretching, angle deformation and dihedral deformations, respectively. Bonded interactions act between bonded sites i, j, k, l with a distance at the equilibrium of d_b, angle ϕ_a and dihedral angles χ_d and with elastic stiffness of K_b, K_a and K_d (for bond, angle and dihedral, respectively). The triple helical geometrical parameters of the collagen triple helix (that is, d_b, ϕ_a and χ_d) are obtained by performing a statistical analysis on a set of five collagen-like molecules available in the Protein Data Bank (identification codes: 2DRT, 1K6F, 1QSU, 1CAG, 1V6Q). The bond lengths between backbone beads, bonding angles and dihedral angles are computed on the basis of these five crystallographic structures, providing a bond reference length (d_b) of 0.3650.07 nm, a bonding reference angle (ϕ_a) of 119.28.72, and a dihedral reference angle (χ_d) of -89.39.76. In order to assess the stiffnesses of bond, angle and dihedrals (respectively, K_b, K_a and K_d), simple atomistic oligopeptides are considered. In order to derive the bonding constant K_b, a glycine-proline structure is chosen. In order to derive the angle elastic constant K_a, a glycine-proline-hydroxyproline oligopeptide is considered. Finally, to obtain the dihedral elastic constants K_d a glycine-proline-hydroxyproline-glycine molecule is studied. These structures are considered both in their atomistic form and in their coarse-grained form, and the parameters of the coarse-grain model (i.e, K_b, K_a and K_d) are identified matching force-extension curves obtained from atomistic and coarse-grained steered molecular dynamics simulations. For the coarse-grained models, the spring constants (that is, the parameters K_b, K_a and K_d) are initially set equal to the values adopted in the original MARTINI force field for an alpha helix secondary structure and then gradually changed in order to best fit the atomistic force-extension curves so that they show the same stiffnesses. The values that best replicate the atomistic behavior are K_b=1250 kJ·mol^{-1}nm^{-2}, K_a=150 kJ·mol^{-1} and K_d=100 kJ·mol^{-1}. In order to validate the extended MARTINI force field model, an 8 nm long collagen-like molecule (similar to the PDB entry used to assess the geometrical features) is pulled under the same conditions previously described. The chosen molecule is [(glycine-proline-hydroxyproline)$_{10}$]$_3$, which represents an ideal reference collagen molecule. The equilibrated coarse-grained collagen peptide is pulled using a deformation rate of 1 m/s and the Youngs modulus of the coarse-grain collagen molecule is then calculated from the slope of the force-deformation plot. Assuming a cylindrical geometry, and considering a collagen molecular radius of 0.8 nm, the Youngs modulus is obtained equal to 4.62±0.41 GPa. The whole structure of the heterotrimeric type I collagen molecule is investigated in order to derive the persistence length. The full length coarse-grained tropocollagen model is divided into five 60 nm peptide strands to reduce the computational cost. Then, the flexibility of the structures is analyzed from molecular dynamics trajectories of 350 ns each, obtaining the persistence length through the following expression:

$$log < cos(\theta) >= -\frac{s}{L_p} \tag{12}$$

As shown in the inlay in fig. 13, the variable θ denotes the angle between two segments along the molecule separated by contour length s (that is, the sum of the segment lengths), and L_p is the persistence length, which provides information about a molecules flexibility. In order to evaluate the persistence length, the molecule is divided into several smaller segments. Each segment (as shown in the inlay in fig. 13), is delimited by the center of mass of the three equivalent glycine residues belonging to the three different collagen chains. The plot of $log < cos(\theta) >$ versus s is obtained, where the average of $cos(\theta)$ is calculated over the full molecular dynamics simulations trajectories. Fig. 13 shows the plot of $log < cos(\theta) >$ versus the contour length s. The value of the persistence length L_p is then obtained from the inverse of the slope after a linear regression of all data, giving a value of L_p = 51.5±6.7 nm.

The persistence length found based on the coarse-grained model is close to the value obtained by Hofmann and coworkers, who found a value of 57±5 nm for collagen type I molecule through an electron microscopy analysis. On the other hand, a literature analysis shows that great variability in persistence length value can be found considering different kinds of experimental tests. The pioneering work in this analysis is represented by Utiyama and coworkers (Utiyama et al., 1973), who considered sedimentation constants and the intrinsic viscosity of purified collagen molecules and

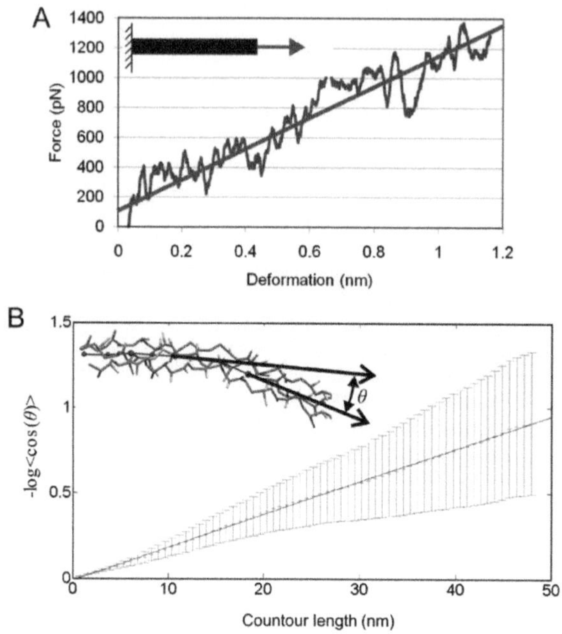

Figure 13. Mechanical analysis of collagen molecules. Panel A: Force-deformation plot up to 15% strain (1.2 nm) for the pulling test of a $[(\text{glycine-proline-hydroxyproline})_{10}]_3$ collagen-like peptide (straight line) and linear regression (dashed line) used to obtain the molecular elastic constant. Panel B: Plot of $log < cos(\theta) >$ versus contour length calculated for the coarse-grain model of the first strand ($\approx$ 60 nm) of human collagen type I molecule. The linear regression of the data for the five different strands leads to an average persistence length of $\approx$51.5 nm for human type I collagen. Adapted from (Gautieri et al., 2010).

calculated a value for the persistence length close to 130 nm. Saito and coworkers (Saito et al., 1982) considering the hydrodynamic properties of collagen derived the intrinsic viscosity and the sedimentation constant of collagen and, from these values, the persistence length, equal to 160-180 nm. In another experiment, Nestler and coworkers (Nestler et al., 1983) measured the dynamic viscoelastic properties of dilute solutions of collagen molecules. From the obtained values of intrinsic viscosity and rotational relaxation time, they found a value for the persistence length of about 170 nm. In more recent work, Sun and co-workers used optical tweezers in order to obtain, from force-extension plots fitted to the Marko-Siggia entropic elastic model, the collagen persistence length, which was found to be 14.5±7.3 nm. Finally, molecular simulation studies were used to perform bending tests on a collagen-like peptide, predicting the bending stiffness of the collagen molecule and, from this result, evaluate the persistence length of collagen to be in the range of 16-24 nm (Buehler and Wong, 2007). The variability that can be found in literature for the collagen persistence length value depends on the used experimental setup, for example electron micrographs or optical tweezers, on the experimental conditions (how the sample is extracted and prepared), on the method applied and the parameters used to fit experimental data with theoretical models. Despite the large variability of experimental values, the value obtained with the coarse grained model is in good agreement with the consideration of collagen as a semiflexible molecule with a rodlike configuration. Due to its contour length of 300 nm, all-atom simulations with explicit solvent are prohibitive since they would require excessive computational resources due to the very large number of particles. The reduction by roughly a factor of 10 in the coarse-grained description provides a significant speedup that facilitates the direct simulation of much longer molecules. Furthermore, the typical time step used in classical molecular dynamics simulations (1-2 fs) allows the modeling only on the nanoseconds time scale. Whereas the coarse-grained model enables to use much longer time-steps, on the order of 20-40 fs. Considering the combined effect of the reduction of the number of particles and the increased time step, the coarse-grained approach leads to a total speed-up of 200-400 with respect to atomistic simulations. With this significant computational speedup, the coarse-grained modeling framework opens many possibilities for future studies, particularly at the scale of collagen fibrils and possibly fibers. However, the MARTINI coarse grain approach is only valid when the phenomena under study do not involve changes of the secondary structure. In the context of collagen, only events that do not involve unfolding of the collagen triple helix can be correctly modeled using the MARTINI force field formulation. While unfolding of molecules is likely to play an important role in large deformation and fracture of collagenous tissues, the mechanical stresses experienced by collagen tissues under small mechanical loads may not lead to unfolding at the molecular level. This is because strain is distributed over several hierarchical levels (*i.e.*, fibers, fibrils, molecules) and involves several concurrent mechanisms (fiber uncrimping, proteoglycan-mediated fibril sliding, molecular slippage and molecular elongation).

9 Conclusions

Recently, - omics approach has become of daily use in biology thanks to inexpensive gene expression profiling techniques and user friendly data-mining software. In the field of material science, a systematic bottom-up knowledge can complement screening methods with the aim of reverse engineering (biological) materials. This is particularly true for complex biological materials, where a holistic perspective is absolutely necessary. With such holistic perspective in mind, all different investigative approaches (experimental, theoretical, computational) can be considered a materiomic approach (Cranford et al., 2013). However, the fundamental challenge is not in the data collection, but rather their combination, which requires the joint effort of biologists, material scientists, chemists, and engineers. The development and combination of both experimental and computational techniques for studying biomaterials will help us to get key understanding of how molecules, cells, tissue and artificial materials interact. Furthermore, if we are able to completely understand complex biological materials, the understanding of synthetic materials should come relatively easier.

The study of material properties at multiple scales is crucial for biological materials, as their structure changes with hierarchical level (and thus length-scale) and, as a consequence, most material properties are strongly dependent on the observation scale. Multi-scale simulation analyses are thus the key to improve our mechanistic understanding of how structure and properties in biological materials are linked. Typically this is achieved from a bottom-up approach, linking more sophisticated lower length-scale parameters, which form the building blocks of the system at that level, to coarser, larger length-scales. Purely bottom-up approaches, though, are incomplete if they lack the link between large-scale behavior to small-scale phenomena. Full stratification of different levels of hierarchy using such analysis develops a powerful feedback loop where the bottom-up modeling approach catalyzes the insights we gain at each layer of the material ladder, with the possibility of controlling properties at multiple scales simultaneously, and to examine their effect on system function. While a materials science approach can probe the physical and chemical components and structure of the system a systematic blueprint the whole is indeed greater than the sum of its parts. Such challenges necessitate a new, holistic and integrated perspective an omic perspective, i.e., materiomics.

Bibliography

B. J. Alder and T. E. Wainwright. Phase transition for a hard sphere system. *Journal of Chemical Physics*, 27(5):1208–1209, 1957.

B. J. Alder and T. E. Wainwright. Studies in molecular dynamics .1. general method. *Journal of Chemical Physics*, 31(2):459–466, 1959.

B. J. Alder and T. E. Wainwright. Studies in molecular dynamics .2. behavior of a small number of elastic spheres. *Journal of Chemical Physics*, 33(5):1439–1451, 1960.

R. M. Alexander. Elastic energy stores in running vertebrates. *American Zoologist*, 24(1):85–94, 1984.

R.M. Alexander. *Animal Mechanics*. Blackwell Scientific, Oxford, UK, 2nd edition edition, 1983.

K. Beck, V. C. Chan, N. Shenoy, A. Kirkpatrick, J. A. M. Ramshaw, and B. Brodsky. Destabilization of osteogenesis imperfecta collagen-like model peptides correlates with the identity of the residue replacing glycine. *P. Natl. Acad. Sci. USA*, 97(8):4273–4278, 2000.

G. I. Bell. Theoretical-models for the specific adhesion of cells to cells or to surfaces. *Advances in Applied Probability*, 12(3):566–567, 1980.

J. Bella, M. Eaton, B. Brodsky, and H. M. Berman. Crystal-structure and molecular-structure of a collagen-like peptide at 1.9-angstrom resolution. *Science*, 266(5182):75–81, 1994.

R. Bhowmik, K. S. Katti, and D. R. Katti. Mechanics of molecular collagen is influenced by hydroxyapatite in natural bone. *Journal of Materials Science*, 42(21):8795–8803, 2007.

R. Bhowmik, K. S. Katti, and D. R. Katti. Mechanisms of load-deformation behavior of molecular collagen in hydroxyapatite-tropocollagen molecular system: Steered molecular dynamics study. *Journal of Engineering Mechanics Ascc*, 135(5):413 421, 2009.

B. Brodsky and J. A. M. Ramshaw. The collagen triple-helix structure. *Matrix Biology*, 15(8-9):545–554, 1997.

M. J. Buehler and S. Keten. Colloquium: Failure of molecules, bones, and the earth itself. *Reviews of Modern Physics*, 82(2):1459–1487, 2010.

M.J. Buehler. Atomistic and continuum modeling of mechanical properties of collagen: Elasticity, fracture and self-assembly. *J. Mater. Res.*, 21(8): 1947–1961, 2006a.

M.J. Buehler. Nature designs tough collagen: Explaining the nanostructure of collagen fibrils. *P. Natl. Acad. Sci. USA*, 103(33):1228512290, 2006b.

M.J. Buehler and S.Y. Wong. Entropic elasticity controls nanomechanics of single tropocollagen molecules. *Biophysical Journal*, 93(1):37–43, 2007.

P. Chamberlain, R. Drewello, L. Korn, W. Bauer, T. Gough, A. Al-Fouzan, M. Collins, N. Van Doorn, O. Craig, and C. Heron. Construction of the khoja zaynuddin mosque: use of animal glue modified with urine. *Archaeometry*, 53:830–841, 2011.

P. Y. Chou and G. D. Fasman. Prediction of protein conformation. *Biochemistry*, 13(2):222–245, 1974.

S. W. Cranford and M. J. Buehler. *Biomateriomics*, volume 165 of *Springer Series in Materials Science*. Springer Netherlands, 2012.

S. W. Cranford, J. de Boer, C. van Blitterswijk, and M. J. Buehler. Materiomics: An -omics approach to biomaterials research. *Adv Mater*, 2013.

J.D. Currey. *Bones: Structure and Mechanics*. Princeton University Press, Princeton, NJ, 2002.

A. A. Deniz, S. Mukhopadhyay, and E. A. Lemke. Single-molecule biophysics: at the interface of biology, physics and chemistry. *Journal of the Royal Society Interface*, 5(18):15–45, 2008.

S. J. Eppell, B. N. Smith, H. Kahn, and R. Ballarini. Nano measurements with micro-devices: mechanical properties of hydrated collagen fibrils. *Journal Of The Royal Society Interface*, 3(6):117–121, 2006.

D. R. Eyre, M. A. Weis, and J. J. Wu. Advances in collagen cross-link analysis. *Methods*, 45:65–74, 2008.

P. Fratzl. *Collagen: Structure and Mechanics*. Springer, 2008.

P. Fratzl and R. Weinkamer. Nature's hierarchical materials. *Progress in Material Science*, 52:1263–1334, 2007.

P. Fratzl, N. Fratzlzelman, and K. Klaushofer. Collagen packing and mineralization - an x-ray-scattering investigation of turkey leg tendon. *Biophysical Journal*, 64(1):260–266, 1993.

P. Fratzl, K. Misof, I. Zizak, G. Rapp, H. Amenitsch, and S. Bernstorff. Fibrillar structure and mechanical properties of collagen. *Journal of Structural Biology*, 122(1-2):119–122, 1998.

Y. C. Fung. Elasticity of soft tissues in simple elongation. *Am J Physiol*, 213(6):1532–44, 1967.

A. Gautieri, S. Vesentini, F. M. Montevecchi, and A. Redaelli. Mechanical properties of physiological and pathological models of collagen peptides investigated via steered molecular dynamics simulations. *Journal of Biomechanics*, 41(14):3073–3077, 2008.

A. Gautieri, M. J. Buehler, and A. Redaelli. Deformation rate controls elasticity and unfolding pathway of single tropocollagen molecules. *Journal of the Mechanical Behavior of Biomedical Materials*, 2(2):130–137, 2009.

A. Gautieri, A. Russo, S. Vesentini, A. Redaelli, and M. J. Buehler. Coarse-grained model of collagen molecules using an extended martini force field. *Journal of Chemical Theory and Computation*, 6(4):1210–1218, 2010.

A. Gautieri, S. Vesentini, A. Redaelli, and M. J. Buehler. Hierarchical structure and nanomechanics of collagen microfibrils from the atomistic scale up. *Nano Letters*, 11(2):757–766, 2011.

A. Gautieri, S. Vesentini, A. Redaelli, and M. J. Buehler. Viscoelastic properties of model segments of collagen molecules. *Matrix Biology*, 31 (2):141–149, 2012.

B. R. Gelin and M. Karplus. Sidechain torsional potentials and motion of amino-acids in proteins - bovine pancreatic trypsin-inhibitor. *Proceedings of the National Academy of Sciences of the United States of America*, 72 (6):2002–2006, 1975.

H. S. Gupta, W. Wagermaier, G. A. Zickler, D. R. B. Aroush, S. S. Funari, P. Roschger, H. D. Wagner, and P. Fratzl. Nanoscale deformation mechanisms in bone. *Nano Letters*, 5(10):2108–2111, 2005.

R. Harley, D. James, A. Miller, and J. W. White. Phonons and the elastic moduli of collagen and muscle. *Nature*, 267(5608):285–7, 1977.

R. C. Haut and R. W. Little. A constitutive equation for collagen fibers. *J Biomech*, 5(5):423–30, 1972.

K. E. Kadler, C. Baldock, J. Bella, and R. P. Boot-Handford. Collagens at a glance. *Journal of Cell Science*, 120(12):1955–1958, 2007.

M. Levitt and A. Warshel. Computer-simulation of protein folding. *Nature*, 253(5494):694–698, 1975.

C. A. Lopez, A. J. Rzepiela, A. H. de Vries, L. Dijkhuizen, P. H. Hunenberger, and S. J. Marrink. Martini coarse-grained force field: Extension to carbohydrates. *Journal of Chemical Theory and Computation*, 5(12): 3195–3210, 2009.

A. C. Lorenzo and E. R. Caffarena. Elastic properties, young's modulus determination and structural stability of the tropocollagen molecule: a computational study by steered molecular dynamics. *Journal Of Biomechanics*, 38(7):1527–1533, 2005.

A. D. Mackerell. Empirical force fields for biological macromolecules: Overview and issues. *Journal of Computational Chemistry*, 25(13):1584–1604, 2004.

A. D. MacKerell, D. Bashford, M. Bellott, R. L. Dunbrack, J. D. Evanseck, M. J. Field, S. Fischer, J. Gao, H. Guo, S. Ha, D. Joseph-McCarthy, L. Kuchnir, K. Kuczera, F. T. K. Lau, C. Mattos, S. Michnick, T. Ngo, D. T. Nguyen, B. Prodhom, W. E. Reiher, B. Roux, M. Schlenkrich, J. C. Smith, R. Stote, J. Straub, M. Watanabe, J. Wiorkiewicz-Kuczera, D. Yin, and M. Karplus. All-atom empirical potential for molecular modeling and dynamics studies of proteins. *Journal of Physical Chemistry B*, 102(18):3586–3616, 1998.

S. J. Marrink, H. J. Risselada, S. Yefimov, D. P. Tieleman, and A. H. de Vries. The martini force field: coarse grained model for biomolecular simulations. *Journal of Physical Chemistry B*, 111(27):7812–24, 2007.

J. A. Mccammon, B. R. Gelin, and M. Karplus. Dynamics of folded proteins. *Nature*, 267(5612):585–590, 1977.

N. Metropolis, A. W. Rosenbluth, M. N. Rosenbluth, A. H. Teller, and E. Teller. Equation of state calculations by fast computing machines. *Journal of Chemical Physics*, 21(6):1087–1092, 1953.

L. Monticelli, S. K. Kandasamy, X. Periole, R. G. Larson, D. P. Tieleman, and S. J. Marrink. The martini coarse-grained force field: Extension to proteins. *Journal of Chemical Theory and Computation*, 4(5):819–834, 2008.

K. Nakajima and T. Nishi. Nanoscience of single polymer chains revealed by nanofishing. *Chemical Record*, 6(5):249–258, 2006.

M. T. Nelson, W. Humphrey, A. Gursoy, A. Dalke, L. V. Kale, R. D. Skeel, and K. Schulten. Namd: A parallel, object oriented molecular dynamics program. *International Journal Of Supercomputer Applications And High Performance Computing*, 10(4):251–268, 1996.

F. H. M. Nestler, S. Hvidt, J. D. Ferry, and A. Veis. Flexibility of collagen determined from dilute-solution viscoelastic measurements. *Biopolymers*, 22(7):1747–1758, 1983.

J.P.R.O. Orgel, T.C. Irving, A. Miller, and T. J. Wess. Microfibrillar structure of type i collagen in situ. *P. Natl. Acad. Sci. USA*, 103(24):9001–9005, 2006.

D. A. Pearlman, D. A. Case, J. W. Caldwell, W. S. Ross, T. E. Cheatham, S. Debolt, D. Ferguson, G. Seibel, and P. Kollman. Amber, a package of computer-programs for applying molecular mechanics, normal-mode analysis, molecular-dynamics and free-energy calculations to simulate the structural and energetic properties of molecules. *Computer Physics Communications*, 91(1-3):1–41, 1995.

E. Pena, J. A. Pena, and M. Doblare. On modelling nonlinear viscoelastic effects in ligaments. *Journal of Biomechanics*, 41(12):2659–2666, 2008.

J. A. Petruska and A. J. Hodge. Subunit model for tropocollagen macromolecule. *Proceedings of the National Academy of Sciences of the United States of America*, 51(5):871–&, 1964.

S. Plimpton. Fast parallel algorithms for short-range molecular-dynamics. *Journal of Computational Physics*, 117(1):1–19, 1995.

J. W. Ponder and D. A. Case. *Force fields for protein simulations*, volume 66 of *Advances In Protein Chemistry*, pages 27–+. 2003.

R. Puxkandl, I. Zizak, O. Paris, J. Keckes, W. Tesch, S. Bernstorff, P. Purslow, and P. Fratzl. Viscoelastic properties of collagen: synchrotron radiation investigations and structural model. *Philosophical*

Transactions Of The Royal Society Of London Series B-Biological Sciences, 357(1418):191–197, 2002.

Z. Qin and M. J. Buehler. Molecular dynamics simulation of the alpha-helix to beta-sheet transition in coiled protein filaments: Evidence for a critical filament length scale. *Physical Review Letters*, 104(19), 2010.

A. Rahman and Stilling.Fh. Molecular dynamics study of liquid water. *Journal of Chemical Physics*, 55(7):3336–&, 1971.

Kartha G. Ramachandran, G.N. Structure of collagen. *Nature*, 176:593595, 1955.

A. K. Rappe, C. J. Casewit, K. S. Colwell, W. A. Goddard, and W. M. Skiff. Uff, a full periodic-table force-field for molecular mechanics and molecular-dynamics simulations. *Journal of the American Chemical Society*, 114(25):10024–10035, 1992.

F. Rauch and F. H. Glorieux. Osteogenesis imperfecta. *The Lancet*, 363: 1377–85, 2004.

A. Redaelli, S. Vesentini, M. Soncini, P. Vena, S. Mantero, and F. M. Montevecchi. Possible role of decorin glycosaminoglycans in fibril to fibril force transfer in relative mature tendons - a computational study from molecular to microstructural level. *Journal Of Biomechanics*, 36(10): 1555–1569, 2003.

A. Rich and F.H.C. Crick. The structure of collagen. *Nature*, 176:915–916, 1955.

B. J. Rigby, N. Hirai, J. D. Spikes, and H. Eyring. The mechanical properties of rat tail tendon. *J Gen Physiol*, 43(2):265–83, 1959.

T. Saito, N. Iso, H. Mizuno, N. Onda, H. Yamato, and H. Odashima. Semi-flexibility of collagens in solution. *Biopolymers*, 21(4):715–728, 1982.

M. S. Sansom, K. A. Scott, and P. J. Bond. Coarse-grained simulation: a high-throughput computational approach to membrane proteins. *Biochemical Society Transactions*, 36(Pt 1):27–32, 2008.

N. Sasaki and S. Odajima. Elongation mechanism of collagen fibrils and force-strain relations of tendon at each level of structural hierarchy. *Journal Of Biomechanics*, 29(9):1131–1136, 1996.

N. Sasaki, N. Shukunami, N. Matsushima, and Y. Izumi. Time-resolved x-ray diffraction from tendon collagen during creep using synchrotron radiation. *Journal of Biomechanics*, 32(3):285–292, 1999.

H. A. Scheraga, M. Khalili, and A. Liwo. Protein-folding dynamics: Overview of molecular simulation techniques. *Annual Review of Physical Chemistry*, 58:57–83, 2007.

H. R. C. Screen, A. Anssari-Benam, and D. L. Bader. A combined experimental and modelling approach to aortic valve viscoelasticity in tensile deformation. *Journal of Materials Science-Materials in Medicine*, 22(2): 253–262, 2011.

Z. L. Shen, M. R. Dodge, H. Kahn, R. Ballarini, and S. J. Eppell. Stress-strain experiments on individual collagen fibrils. *Biophysical Journal*, 95 (8):3956–3963, 2008.

Z. L. Shen, H. Kahn, R. Ballarini, and S. J. Eppell. Viscoelastic properties of isolated collagen fibrils. *Biophys J*, 100(12):3008–15, 2011.

Z. L. L. Shen, M. R. Dodge, H. Kahn, R. Ballarini, and S. J. Eppell. In vitro fracture testing of submicron diameter collagen fibril specimens. *Biophysical Journal*, 99(6):1986–1995, 2010.

F. H. Silver, D. L. Christiansen, P. B. Snowhill, and Y. Chen. Transition from viscous to elastic-based dependency of mechanical properties of self-assembled type i collagen fibers. *Journal of Applied Polymer Science*, 79 (1):134–142, 2001.

M. Srinivasan, S. G. M. Uzel, A. Gautieri, S. Keten, and M. J. Buehler. Alport syndrome mutations in type iv tropocollagen alter molecular structure and nanomechanical properties. *Journal of Structural Biology*, 168 (3):503–510, 2009.

Y. L. Sun, Z. P. Luo, A. Fertala, and K. N. An. Direct quantification of the flexibility of type i collagen monomer. *Biochemical And Biophysical Research Communications*, 295(2):382–386, 2002.

R. B. Svensson, T. Hassenkam, P. Hansen, and S. P. Magnusson. Viscoelastic behavior of discrete human collagen fibrils. *Journal of the Mechanical Behavior of Biomedical Materials*, 3(1):112–115, 2010a.

Ren B. Svensson, Tue Hassenkam, Colin A. Grant, and S. Peter Magnusson. Tensile properties of human collagen fibrils and fascicles are insensitive to environmental salts. *Biophysical Journal*, 99(12):4020–4027, 2010b.

V. Tozzini. Coarse-grained models for proteins. *Current Opinion in Structural Biology*, 15(2):144–150, 2005.

H. Utiyama, K. Sakato, K. Ikehara, T. Setsuiye, and M. Kurata. Flexibility of tropocollagen from sedimentation and viscosity. *Biopolymers*, 12(1): 53–64, 1973.

Joost A. J. van der Rijt, Kees O. van der Werf, Martin L. Bennink, Pieter J. Dijkstra, and Jan Feijen. Micromechanical testing of individual collagen fibrils. *Macromolecular Bioscience*, 6(9):697–702, 2006.

D. Van der Spoel, E. Lindahl, B. Hess, G. Groenhof, A. E. Mark, and H. J. C. Berendsen. Gromacs: Fast, flexible, and free. *Journal of Computational Chemistry*, 26(16):1701–1718, 2005.

W. Wang, O. Donini, C. M. Reyes, and P. A. Kollman. Biomolecular simulations: Recent developments in force fields, simulations of enzyme catalysis, protein-ligand, protein-protein, and protein-nucleic acid noncovalent interactions. *Annual Review of Biophysics and Biomolecular Structure*, 30:211–243, 2001.

X. T. Wang and R. F. Ker. Creep-rupture of wallaby tail tendons. *Journal of Experimental Biology*, 198(3):831–845, 1995.

S. Weiner and H. D. Wagner. The material bone: Structure mechanical function relations. *Annual Review Of Materials Science*, 28:271–298, 1998.

Microelectromechanical systems (MEMS) platforms for testing the mechanical properties of collagen fibrils

Roberto Ballarini [*] and Harold Kahn [†]

[*] Department of Civil Engineering, University of Minnesota, Minneapolis, USA

[†] Department of Materials Science and Engineering, Case Western Reserve University, Cleveland, USA

1 Introduction

To determine how the hierarchical structures of tissues such as skin, tendon and bone confer mechanical properties to those tissues a better understanding is required of the properties of the individual components and their interactions. The architecture of bone, for example, contains six distinct features shown in Figure 1 (Rho et al. (1998)). The building blocks are collagen molecules and mineral crystals. The figure illustrates how the feature at a particular scale is assembled from those at the lower scales. The ultimate goal of computational and theoretical multiscale models is the accurate description of the properties and mechanical behavior of such complex structures. The development of robust models will undoubtedly rely on experimental data that (1) will provide some of the required input parameters; (2) could be used to assess the validity of the model's predictions; and (3) will offer insights on how they could be improved. The mechanical testing of biological tissues at relatively large scales enjoys a long and rich history. However, data on the micro and nano scale features is sparse at best because of the difficulties associated with specimen preparation and handling and with the accurate measurement of forces and displacements. The mechanical properties of the larger features of bone, for example, has been studied using standard (albeit often customized) testing equipment, and more recently optical techniques including those described in Chapter 5 of this book have enabled the probing of the structural response of the molecular scale components. Until recently there was a void in experimentation of collagen fibrils and fibers because testing platforms were not available to measure forces and displacements to the appropriate dimensions. The void was partially filled by the development of microelectromechanical systems

(MEMS) devices that produced first-time measurements of the strength, strain capacity, and viscoelastic properties of collagen fibrils subjected to simple elongation. This chapter provides a brief summary of the MEMS approach to mechanical testing and experimental data that illustrates the complex response of collagen fibrils to mechanical forces.

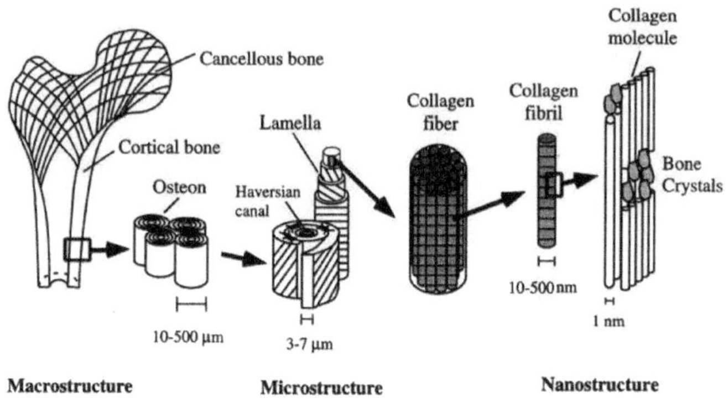

Figure 1. Hierarchical structure of bone (Rho et al. (1998)).

2 MEMS structural testing platforms

The authors and their coworkers designed, fabricated and used two generations of MEMS structural testing platforms to measure the force-displacement-time response of collagen fibrils subjected to simple elongation (Eppell et al. (2006); Shen et al. (2008, 2010, 2011)). The two generations differed in the way that the force was applied and the displacement was measured. The evolution and details of the MEMS technology used to design and fabricate the testing platforms, specimen preparation procedures, and structural testing protocols can be recovered from the references. The first generation devices, which are not described in any detail in this chapter, used electrostatic actuation to apply force to the specimen, and the displacements were measured using a Vernier scale. The second generation devices rely on piezoelectric displacement transducers capable of applying displacement-controlled motions with nm-scale resolution. Digital-image-correlation software in conjunction with in-situ microscopy with digital image capture enabled displacement measurements in real-time at nm-scales. Representative data obtained from experiments using the two different devices is summarized in the next section. For both generation devices accurate determination of force from measured displacement was accomplished by using standard MEMS technology, as follows.

The basic idea of the MEMS paradigm is that if displacements of an elastic test structure with known dimensions and elastic moduli can be measured accurately, the forces can be determined using a structural analysis. One common material used to create MEMS devices is polycrystalline silicon. Standard MEMS processing techniques of deposition, patterning, and etching can produce polycrystalline silicon structures with dimensions ranging from sub-micrometer to hundreds of micrometers. The elastic properties of silicon are well-known but slightly anisotropic. Therefore the structural analyses of devices made from single-crystal silicon or large-grained polycrystalline silicon can be significantly more complicated than those performed for homogeneous isotropic structures. Fortunately, polycrystalline silicon deposited by chemical vapour deposition at temperatures below 600^oC contains grain sizes on the order of 100 nm (Kahn et al. (2002)). Therefore, typical size MEMS devices will contain large numbers of silicon grains in all in-plane dimensions, leading to nearly isotropic in-plane elastic moduli (Mullen et al. (1997)) .

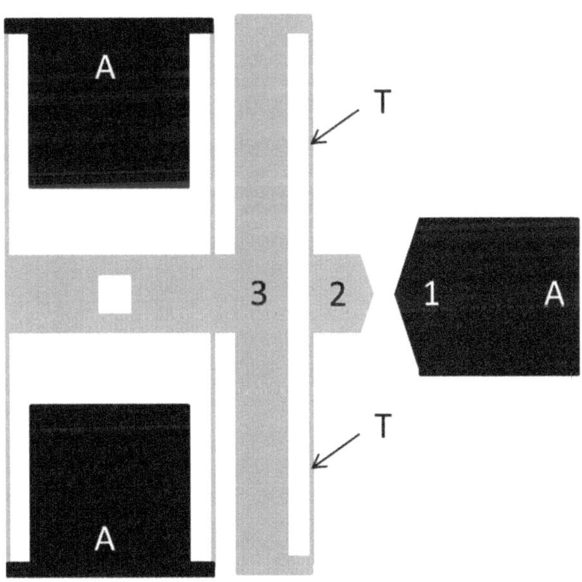

Figure 2. Schematic of MEMS platform for testing collagen fibrils and other fiber-like structures.

The second generation device used most recently to test individual collagen fibrils (Shen et al. (2010, 2011)) is shown schematically in Figure 2. Multiple devices were fabricated with standard MEMS processing techniques, using 100 mm diameter silicon wafers as substrates. Using low-pressure chemical vapor deposition the wafers were sequentially coated with 3.5 micrometers of sacrificial silicon dioxide, and 6 micrometers of polycrystalline Si. The deposited layers were sufficiently annealed to relieve any growth-induced residual stresses. Devices were patterned using optical lithography, and then the polycrystalline silicon was etched in a sulphur hexafluoride plasma. The photoresist was removed in aqueous sulphuric acid and hydrogen peroxide, and devices were released by immersion in aqueous hydrofluoric acid, followed by rinsing in methanol and critical point carbon dioxide drying. The release etch dissolved all the sacrificial silicon dioxide beneath the portions of the device that were designed to be movable. The majority of the sacrificial silicon dioxide was left beneath areas of the device designed as anchors in order to keep the device attached to the silicon substrate. Because the device is less than $1mm^2$ in size, thousands of devices were fabricated simultaneously on a single silicon wafer.

With respect to the shematic shown in Figure 2, the black areas labeled "A" are anchored to the substrate, while the gray areas are free to move. The device is designed for one end of the collagen fibril to be attached to the pad labeled "1" and the other end to be attached to the pad labeled "2". A scanning electron microscope image of a representative device, upon which a collagen fibril is attached, is shown in Figure 3. The central gray portion of the device is then attached to an external displacement transducer. In the work described in Shen et al. (2010, 2011) this was achieved by placing a needle tip into the square hole created in the device and shown in the schematic and in the image, and by connecting the needle to a piezoelectric transducer-controlled stage. When the device is displaced to the left, as oriented in Figure 2, the movement of all three positions, labeled "1", "2", and "3", must be recorded simultaneously.

Since pad 1 is anchored to the substrate, the elongation of the collagen fibril specimen will be equal to the increase in separation between pads 1 and 2. The force applied to move pad 2 to the left is equal to the force applied to move pad 3 to the left. Therefore, the force applied to the specimen can be determined by measuring the elastic deformation involved in the increase in separation between pads 2 and 3. That separation is controlled by the elastic response of the polycrystalline silicon tether beams labeled "T" in Figure 2. The dimensions of the tether beams can be measured very accurately using electron microscopy or other techniques, and the elastic modulus of polycrystalline silicon is known (Mullen et al. (1997); Jensen

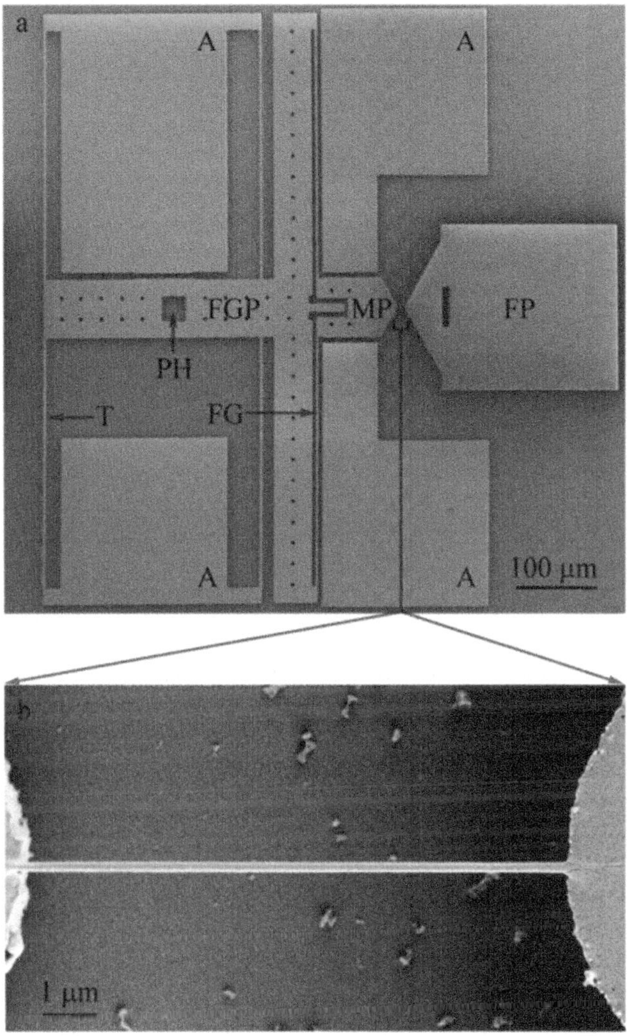

Figure 3. MEMS platform for testing collagen fibrils (Shen et al. (2010, 2011)).

et al. (2001)). This information can be used as input into a finite element model of the device to infer the forces necessary to achieve the observed displacements. In this manner, both the force and elongation of the fiber are determined simultaneously.

Note that the range of available measurements will be optimized if the length, width and thickness of the tether beams are designed to have nearly the same stiffness as the nanoscale specimen. Materials with different stiffnesses can be tested by fabricating many devices with different tether beam stiffnesses on a single silicon wafer substrate, and then using whichever device best matches the material of interest.

3 Experimental results

The collagen fibrils studied using the first generation devices are considered as having been tested in air. This is because the testing machines involve an electrostatic actuator, and therefore could not be immersed in the liquid environment required to maintain the fibrils at 100% humidity throughout the test. Fibrils tested using these devices were pulled out of solution and kept in wet paper in an effort to minimize drying during the time required to attach them to the MEMS platforms and to perform the experiments. The second generation devices, however, enabled the fibrils to be maintained in a liquid environment provided by a liquid drop placed onto the MEMS platform for the duration of the experiment. This section presents representative results that highlight the qualitative and quantitative response of individual collagen fibrils to three types of loading; monotonically increasing tension up to failure; loading to a fraction of the nominal strength followed by repeated unloading and reloading cycles; and creep-relaxation experiments.

3.1 Tensile strength and ultimate strain capacity

A representative nominal stress-nominal strain curve in the hydrated state reproduced from Shen et al. (2010) is shown in Figure 4. Note that the strain is averaged over the length of the fibril and is calculated as the stretch of the fibril divided by the original gage length. The stress, defined as the force divided by the average of cross-sectional area measurements imaged along a finite number of points along the fibril length, does not account for the significantly nonuniform cross-sectional area along the fibril and how it changes throughout the test. For these reasons the true stress and the maximum strain achieved at a particular point along the fibril are expected to be significantly higher than the nominal values. Figure 4 shows that the response is linear and that the failure is brittle; there is no post-peak response for this nearly displacement controlled test. As expected the strength, strain capacity and elastic modulus data presented in Shen et al. (2010) and reproduced in Table 1 showed significant specimen-to-specimen variation. It is remarkable that the fibrils are capable of stretching by as much as $\sim 180\%$ and that their nominal strength can be as high as $0.5GPa$. The large batch-to-batch variation of the response in simple tension suggests that it will prove difficult to assess the validity of multiscale models of collagen fibrils.

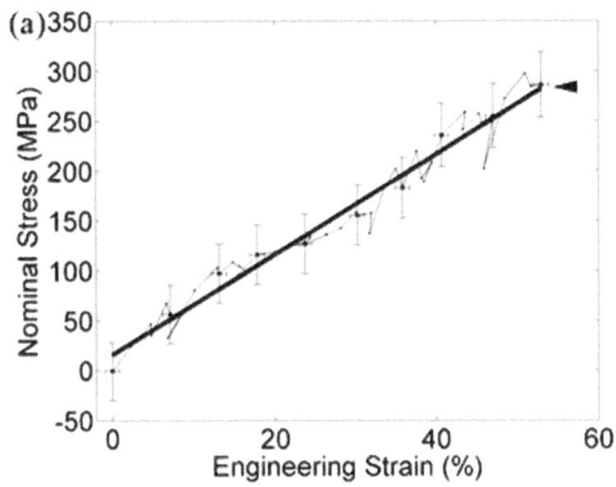

Figure 4. Stress-strain curve of representative collagen fibril up to failure (Shen et al. (2010)).

Table 1. Tensile properties of hydrated collagen fibrils (Shen et al. (2010)).

	Strength (MPa)	Failure strain	Elastic modulus (MPa)
(one standard deviation)	230 ± 160	0.80 ± 0.44	470 ± 410
(range)	40 ± 490	$0.33 - 1.83$	$110 - 1470$

3.2 Initial evidence of viscoelasticity and strain recovery

The first generation MEMS platforms were used to study the response of collagen fibrils to loading/unloading cycles (Shen et al. (2008)). A representative cyclic stress-strain curve is shown in Figure 5, where the unloading and subsequent reloading traces represent the average of the measurements over four cycles. The results show that collagen fibrils tested in air exhibit viscoelasticity and a residual strain upon removal of the load. These observations challenge multiscale models to account for time and possibly rate effects. But this is not the only rich constitutive behavior that was observed in the study reported in Shen et al. (2008). It was discovered that if a fibril is placed in a humid environment at the end of the cyclic loading experiment for a period of approximately one hour, the residual strain completely disappears. This is evidenced by Figure 6, which shows the loading traces obtained from three tests on the same representative fibril discussed above (the top plot corresponds to Figure 5). Each of the two additional tests were conducted after a one hour-long rest period. (Note that the unloading points were removed from the plots to render the figure less busy). It is observed that all stress-strain curves start at the origin, indicating complete strain recovery during the rest period and a modest reduction in stiffness. So in addition to time and possible rate effects, multiscale models must be challenged to account for the fibril's ability to recover strain when subjected to loading and unloading scenarios.

3.3 Creep-relaxation experiments and determination of relaxation times

The time effects observed in the initial experiments inspired the creep-relaxation tests described in Shen et al. (2011). Because of the mechanics of the test platform, the experiments cannot be performed under strictly constant force or constant displacement conditions. However, proper accounting of the force-displacement-time response of the collagen fibrils produced the representative stress-time and strain-time plots shown in Figure 7. This data was fitted well by the Maxwell-Weichert model illustrated in Figure 8, which is associated with three elastic moduli and two viscous elements. The relaxation function associated with the Maxwell-Weicher model is given by

$$E_{relax}\left(t\right) = E_o + E_1 exp\left(-\frac{t}{\tau_1}\right) + E_2\left(-\frac{t}{\tau_2}\right) \qquad (1)$$

where E_o is the time-independent elastic modulus, and the relaxation time of each element, τ_i, is equal to the coefficient of viscosity of its dashpot, η_i, divided by the elastic modulus of its spring, E_i. Figure 9 shows typical fits of the relaxation modulus, and Figure 10, reproduced from Shen et al. (2011), lists the data derived from all the experiments. The results indicate that collagen fibrils have fast and slow time responses of the order of 10 and 100 seconds, respectively. The mechanism of viscoelasticity, discussed in much more detail in Shen et al. (2011), was attributed to molecular rearrangement of collagen and water molecules. A comparison with other

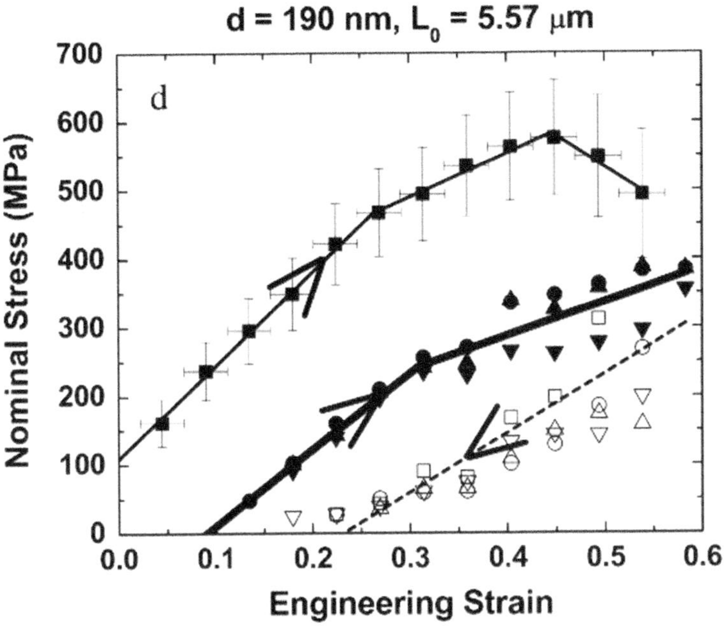

Figure 5. Cyclic loading and unloading of a collagen fibril. The unloading and subsequent loading traces represent the average of four cycles (Shen et al. (2008)).

collagenous structures at larger scales (tissues) and smaller scales (collagen molecules) is warranted. Using Fung's QLV model (Fung (1967); Fung et al.), sheep flexor tendons were shown to have a short relaxation time of $\sim 2s$ and a long relaxation time of $\sim 1500s$ (Sarver et al. (2003)). In a similar way, goat femur medial-collateral-ligament-tibia complexes were shown to have a short relaxation time of $\sim 2s$ and a long relaxation time of $\sim 2200s$ (Abramowitch and Woo (2004)). Thus, compared to the long relaxation time of soft tissues (1500–2200s), the long relaxation time of a single fibril is one order of magnitude smaller. The short relaxation time of the single isolated fibril is, conversely, three to four times longer than the bulk tissue. This is difficult to explain using first principles and may be an artifact of the phenomenological fitting. Another possibility is that the differences in time behavior arise from differences in tissue source. Given the data in hand at this point and assuming that model type and tissue source are not sufficient to explain an order-of-magnitude difference in relaxation

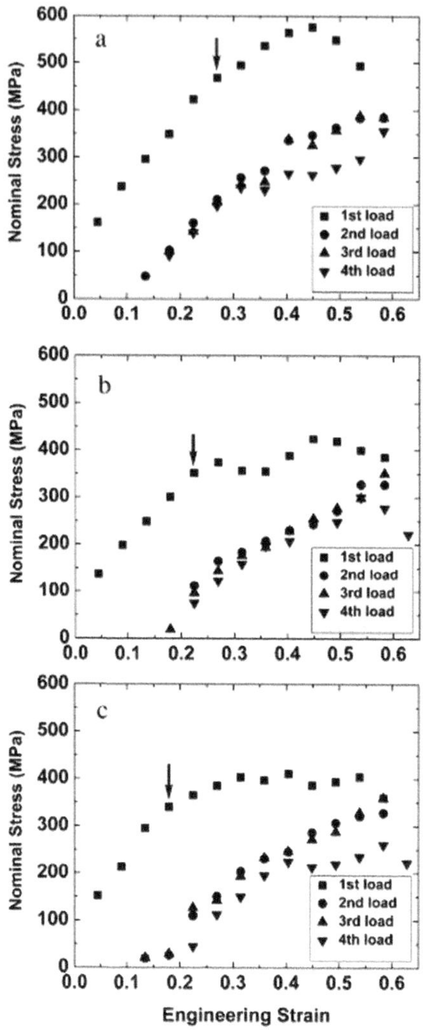

Figure 6. Stress-strain curves produced by three tests on the same fibril with rest periods between the second and third experiments (Shen et al. (2008)).

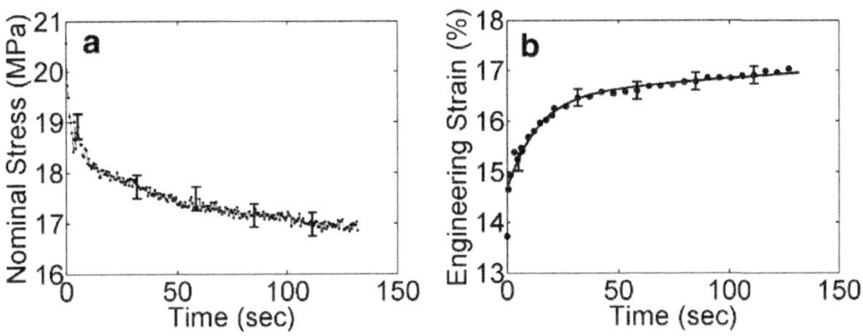

Figure 7. Stress-time and strain-time data of representative collagen fibril (Shen et al. (2011)).

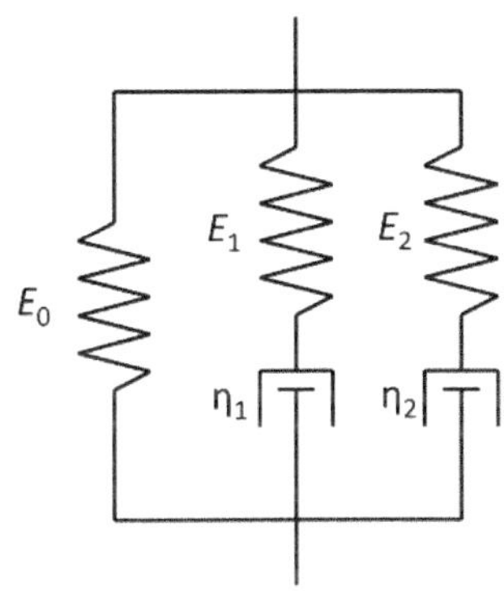

Figure 8. Maxwell-Weichert model.

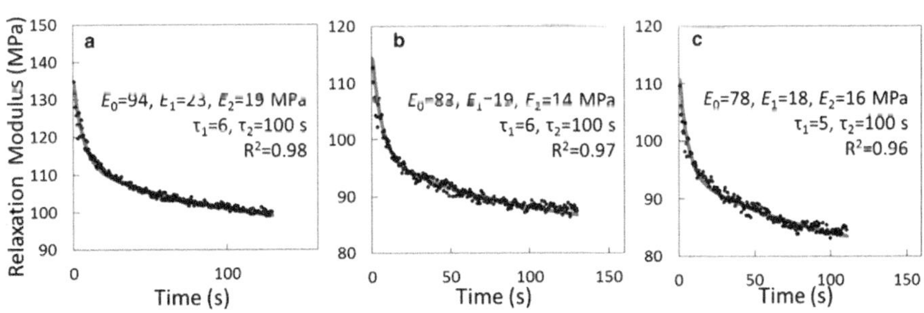

Figure 9. Relaxation modulus of representative fibril as derived using the Maxwell-Weichert model (Shen et al. (2011)).

	E_0 (MPa)	E_1 (MPa)	E_2 (MPa)	τ_1 (s)	τ_2 (s)	R^2
Test 1 (N = 8)	140 ± 50 (80–250)	18 ± 4 (11–23)	20 ± 9 (8–35)	8.1 ± 2.0 (6–12)	100 ± 5 (100–110)	0.95 ± 0.03 (0.90–0.98)
Test 2 (N = 8)	120 ± 40 (70–210)	11 ± 6 (6–22)	13 ± 4 (7–17)	7.1 ± 3.2 (4–13)	100 ± 0 (100–100)	0.91 ± 0.05 (0.82–0.96)
Test 3 (N = 8)	110 ± 40 (60–190)	10 ± 4 (6–18)	13 ± 3 (8–18)	6.6 ± 1.3 (5–9)	100 ± 6 (90–110)	0.92 ± 0.05 (0.83–0.96)
Mean of all tests	123 ± 46	13 ± 6	16 ± 7	7 ± 2	102 ± 5	0.93 ± 0.04

Figure 10. Parameters used to fit viscoelastic response of collagen fibrils using the Maxwell-Weichert model (Shen et al. (2011)).

time, the results suggest that the relaxation times of whole tissues are determined by components other than the collagen fibrils, most likely the ground substance (i.e., proteoglycans). The elastic moduli associated with the Weichert model were compared to those measured during the short-time simple elongation tests. The time-independent elastic modulus of the fibril specimens measured on the first pull was 140 ± 50 MPa, (range 80–250 MPa). This fell within but at the lower end of the 110-1470 MPa range obtained from in vitro monotonic strain-to-fracture tests and listed in Table 1. It is noted that the value of 140 MPa also agrees within a factor of 3 with a recent multiscale model of collagen fibril mechanics Gautieri et al. (2011).

4 Summary

It has become abundantly clear that MEMS devices are ideal testing machines for nanoscale fiber-shaped materials and structures. They enable textbook simple elongation, creep/relaxation, and cyclic loading experiments to be performed with precise measurement of force and displacements, and they can produce data that is instrumental for the development of multiscale models of biological structures. The initial data obtained using the MEMS paradigm highlights the rich constitutive behavior of collagen fibrils, and significant specime-to-specimen scatter. The nonlinear, time dependent and statistical material properties of collagen fibrils therefore provide significant challenges to multiscale modelers.

Bibliography

S.D. Abramowitch and S.L. Woo. An improved method to analyze the stress relaxation of ligaments following a finite ramp time based on the quasi-linear viscoelastic theory. *J. Biomech. Eng.*, 126:92–97, 2004.

S. Eppell, B. Smith, H. Kahn, and R. Ballarini. Anano measurements with micro devices: Mechanical properties of hydrated collagen fibrils. *Journal of the Royal Society Interface*, 3:117–121, 2006.

Y.C. Fung. Elasticity of soft tissues in simple elongation. *Am. J. Physiol.*, 213:1532–1544, 1967.

Y.C. Fung, N. Perrone, and M. Anliker. In *Biomechanics: Its Foundations and Objectives.*

A. Gautieri, S. Vesentini, and M.J. Buehler. Hierarchical structure and nanomechanics of collagen microfibrils from the atomistic scale up. *Nano Lett.*, 11:757–766, 2011.

B.D. Jensen, M.P. de Boer, N.D. Masters, F. Bitsie, and D.A. LaVan. Interferometry of actuated microcantilevers to determine material properties and test structure nonidealities in mems. *Journal of Microelectromechanical Systems*, 10:336–346, 2001.

H. Kahn, A.Q. He, and A.H. Heuer. Homogeneous nucleation during crystallization of amorphous silicon produced by low-pressure chemical vapour deposition. *Philosophical Magazine*, A82:137–165, 2002.

R.L. Mullen, R. Ballarini, and Y. Yin. Monte-carlo simulation of effective elastic constants of polycrystalline thin films. *Acta Metallurgica et Materialia*, 45(6):2247–2255, 1997.

J.Y. Rho, L. Kuhn-Spearing, and P. Zioupos. Mechanical properties and the hierarchical structure of bone. *Med. Eng. Phys.*, 20:92–102, 1998.

J.J. Sarver, P.S. Robinson, and D.M. Elliott. Methods for quasilinear viscoelastic modeling of soft tissue: Application to incremental stress-relaxation experiments. *J. Biomech. Eng.*, 125:754–758, 2003.

Z.L. Shen, M.R. Hodge, H. Kahn, R. Ballarini, and S.J. Eppell. Sress-strain experiments on individual collagen fibrils. *Biopysical Journal*, 95: 3956–3963, 2008.

Z.L. Shen, M.R. Hodge, H. Kahn, R. Ballarini, and S.J. Eppell. In-vitro fracture testing of submicron diameter collagen fibrils. *Biophysical Journal*, 99:1986–1995, 2010.

Z.L. Shen, H. Kahn, R. Ballarini, and S.J. Eppell. Viscoelastic properties of isolated collagen fibrils. *Biophysical Journal*, 100:3008–3015, 2011.

Multiscale modeling of diffusion phenomena in polymers

Alberto Redaelli* Simone Vesentini * Alfonso Gautieri* Paolo Zunino†
* Department of Bioengineerinig, Politecnico di Milano, Milano, Italy
† Department of Mechanical Engineering and Materials Science, University of Pittsburgh, Pittsburgh, USA

Abstract In the present Chapter we will present two examples where the molecular description of diffusion is used to obtain mesoscale behavior of a given bulk polymer. In the first example a large molecule (benzene) is considered as permeant molecule within a bulk polymer (Poly vinyl alcohol), in this case the problem to be overcome was to reach the normal diffusion regime, hence a complete random motion of the diffusive molecules within the polymer. In the second example the diffusing molecules are water and the polymer is poly lattic acid, in this case the problem was to model properly the degradation in time observed when this polymer is exposed to water.

1 Introduction to diffusion at the nanoscale

Together with convection and advection, diffusion is another mechanism by which matter moves. Compared with the two other mechanisms diffusion is chaotic and random, that is there is no preferred direction of motion other than from areas of high to low concentration. This is due to the fact diffusion is a transport phenomenon that does not require bulk motion and the motion is driven by concentration gradient solely. In general, diffusion is a consequence of the constant thermal motion of molecules and particles. There are two ways to describe diffusion: a phenomenological approach dealing with Ficks laws or an atomistic one by considering the random walk of the diffusing particles. In this latter regard diffusion is governed and affected by by both permeant and bulk features and their interactions. From this side, diffusion is considered as a result of the random walk of the diffusing particles self-propelled by thermal energy. The theory of this motion and the atomistic backgrounds of diffusion were developed by Albert Einstein. In 1905 he published in the Annalen der Physik the *annus mirabilis* papers: four articles which contributed substantially to the foundation of modern physics and changed views on space, time, and matter. In the field of diffusion, applying the kinetic theory of fluids he derived expressions for the mean squared displacement of particles, giving a formulation of the diffusion coefficient for a permeant molecule (D) by proposing a statistical mechanics approach to the Brownian motion (Einstein, 1905). In a current perspective Molecular Dynamics (MD) simulations can provide in principle the direct evaluation of the time evolution of a system of interacting particles and can thus be very useful for the determination of D (Hofmann et al., 2009). The major drawback of this method is the limited time scale and sample size that can be simulated. Computer power now available enables to simulate samples up to 10 nm into tens nanosecond domain. Thus the range of properties that can be studied directly is limited to those evolving over this time scale and sample size. The diffusion coefficient D can be directly calculated from the motion of the particles extracted during a MD simulation, but this equation holds only in the case that the observation time (i.e., the simulation time) is large enough to allow the particles to show uncorrelated motion (Hofmann et al. (2009); Entrialgo-Castano et al. (2006); Lendlein and Langer (2002)). In order to assess an accurate diffusion coefficient, MD simulations must reach the normal diffusive regime which, depending on the bulk and the diffusive molecule, can be on time scales higher than nanoseconds. This regime is therefore often difficult to reach for molecules larger than diatomic molecules. Another drawback of such a description of diffusion regards the type of interaction between diffusive molecule and the bulk. In this context classical MD simulations are not able to simulate, for example, covalent bond break as occurring in biodegradable materials and in the cases where the diffusion of permeant molecules within the bulk determines also a molecular mass reduction by bond breaking MD simulations are not useful per se. Nonetheless there is a considerable interest in the study of the transport of molecules across bulk materials, mainly because of the large number of applications in which this transport process plays a major role. These applications include coatings, shape-memory polymeric materials, drug delivery systems, scaffold for tissue engineering and biodegradable materials (Lendlein and Langer (2002); Agrawal and Ray (2001); Soares and Zunino (2010)). In general bulk materials for these applications are polymers due to their versatility and the industrial interest has stimulated the development of theoretical models to describe the transport process across polymeric membranes. There are a large number of models, all of which however lack a correct microscopic description of the permeation process (Soares and Zunino, 2010). Computer simulation methods and in particular MD simulations combined with coarse grain models can become an essential tool to obtain a more detailed picture. A qualitative description of the underlying processes and eventually quantitative predictions of diffusivity open the prospect for the design of membranes with predefined properties.

2 Development of an atomistic informed Coarse Grain model for the study of the diffusion of large permeant molecules within a polymer

Atomistic simulation have been successfully applied in the past to obtain the diffusion coefficients of small molecules (like oxygen, carbon dioxide or water) in polymeric membranes (Gestoso and Karayiannis, 2008; Pavel and Shanks, 2003; Chiessi et al., 2007), polymeric blends Pavel and Shanks (2005), biopolymers Prathab and Aminabhavi (2007). However, despite the increasing computational power available to researchers and the improvements in the MD codes, atomistic simulations are still able to handle only systems with tens or hundreds of thousands of atoms and in the nanoseconds time scale. Several phenomena of interest at the material scale, however, cover time and space scales larger than those affordable with full atom modeling. Recently, the use of coarse grain (CG) modeling, in which a number of atoms is condensed into beads or interacting particles, has proven to be a suitable option to model large systems and long time scales. CG models have been developed whit particular focus on biomolecular systems, since biomolecules are often too large and their characteristic times too long to be treated with full atomistic simulations. However, there are no theoretical impediments to the application of CG methods to analyze also the diffusion behavior. In this view, of particular interest is the CG force field developed by Marrink and co-workers and called MARTINI force field Marrink et al. (2004). The philosophy of the MARTINI force field is to accurately parameterize the basic building blocks of the system thus allowing a broad spectrum of applications (Tozzini, 2005; Chng and Yang, 2008). Concerning diffusion it is possible to build an atomistic-informed parameterization of the MARTINI force field for the modeling of penetrants diffusion in polymeric membranes. Atomistic simulations are used to calculate the interaction free energy between the basic building blocks of the system, i.e. the penetrant molecule and the polymer monomers, and then CG molecular dynamics simulations are performed to assess the diffusion behavior. As an application, we investigated the diffusion behavior of benzene in a bulk of Poly (vinyl alcohol), PVA Gautieri et al. (2010a). The benzene molecule was represented by one bead (or super-atom) and the vinyl alcohol (VA) monomers are represented by a different bead (Figure 6).

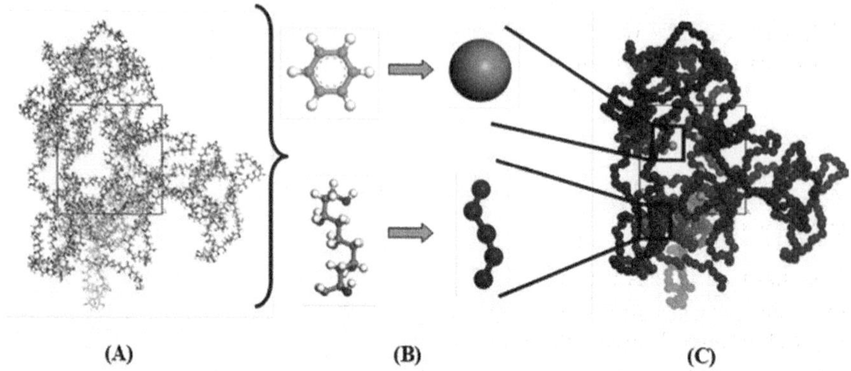

Figure 1. Coarse Grain approach applied on a system composed by benzene and a PVA chain. The CG approach applied on a molecular bulk model (left) by defining superatoms (center) in order to reduce the complexity of the system (right).

The original mapping scheme is based on the assumption that four heavy (non-hydrogen) atoms are grouped into a single bead. Accordingly, benzene molecule is represented by one bead and the VA monomers are represented by a different bead (see Figure 6). The mass of the two types of beads are calculated as the sum of the masses of the atoms grouped into the bead. The chemically bonded beads interact through a bond interaction modeled as a harmonic potential $V_{bond}(r)$:

$$V_{bond}(r) = \frac{1}{2}k_{bond}(r - r_0) \tag{1}$$

where r is the distance between two bonded beads, k_{bond} is the force constant of the bond interaction and r_0 is the equilibrium distance, which have to be obtained by means of atomistic simulations. The beads i and j that are not chemically connected are described by a Lennard-Jones potential (2):

$$V_{LJ}(r) = \epsilon_{ij}\left[\left(\frac{r_{0ij}}{r}\right)^{12} - 2\left(\frac{r_{0ij}}{r}\right)^{6}\right] \tag{2}$$

where$r$ is the distance between the two nonbonded beads, ϵ_{ij} is the energy minimum depth (the strength of the interaction) and r_{0ij} is the distance at the minimum of the potential. We can calculate one set of bonded parameters (for bonded VA-VA beads) and three sets of nonbonded parameters (VA-VA, benzene-benzene and VA-benzene). For each set we run a full atomistic simulation of the two group involved in the interaction and the free energy of interaction between the two groups can be calculated. Eventually, two different molecular models were generated: pure benzene and PVA with a small amount of benzene as penetrant molecule. The two atomistic systems are then converted into CG systems. CG molecular dynamics simulations are carried using the MARTINI force field, which was modified including the parameters for bonded and nonbonded interactions between benzene and VA beads, as obtained from the atomistic simulations. The diffusion coefficient D_i of a single permeant molecule i can be calculated by the Einstein relation, starting from the diffusion trajectory, which is determined during the production simulations, as in full atoms MD simulations.

3 Development of an atomistic informed mesoscale model of a degradable polymer

Polymer degradation is the chain scission process that breaks polymer chains down to oligomers and finally to monomers, ultimately resulting in a decrease of molecular weight. Polymers degrade by several different mechanisms, depending on their inherent chemical structure and on the environment conditions to which they are subjected. The prevailing mechanism of biological degradation is scission of the hydrolytically unstable backbone chain by passive hydrolysis Tamada and Langer (1993). As a consequence of that, the mechanical and transport properties of such systems evolve with time according to the change of relevant indicators such as the water content of the system, i.e. the partial density of water into the water/polymer mixture Soares and Zunino (2010). Being able to study the degradation process is a relevant technological issue. This is important not only to model the degradation of give polymer per se, but also to design the kinetic of a biodegradable drug delivery system. Three distinct but related phenomena occur concurrently:
1. water uptake occurs into the initially dry and non-degraded polymeric network through the mechanism of diffusion;
2. as water becomes readily available in the proximity of the chemical bonds, the likelihood of hydrolytic scission increases, leading to an overall molecular weight reduction;
3. smaller chains are progressively produced, which are more eager to dissolve and diffuse through the network leading to polymer erosion.
In this scenario water diffusion throughout the polymer matrix and polymer weight reduction are key factors which drive the polymer erosion Tamada and Langer (1993). Unfortunately bond rupture is an unaffordable event to

be simulated by standard MD methods. Despite this, the analysis of a succession of MD structures would be very informative. Using a MD approach, it is possible to analyze the diffusion of water, matrices with different degrees of degradation. The degradation of the polymer matrix can be modeled considering monodisperse systems and reducing the length of the polymer chains. The atomistic systems can then be simulated for time scales in the order of the nanoseconds during which the trajectories of water molecules are followed. In particular for a given partial density of water ($\rho_W(t,x)$) within the bulk, the water diffusion (D_w) in a nanoscale element with the same partial density can be evaluated by MD simulations. To exchange and compare this information with macroscopic data, coming for instance from the macroscale model or from any experimental activity, nude MD simulations require to be complemented with suitable downscaling and upscaling operators. In this way the water/polymer mixture density can be obtained; significant outputs at the macroscale are then calculated such as the partial density of water ($\rho_W(t,x)$ (where x denotes the spatial coordinate along the thickness of a polymeric substrate) and the total density of polymer, that is $\rho(t,x) = \sum_{i=1}^{N} \rho_i(t,x)$ where N represents the degree of polymerization of the polymer. The characteristic space and time scales of this analysis are ones relative to the entire polymeric bulk. This is a macroscopic information, where quantities evaluated at point x and time t represent spatial and temporal averages of microscopic phenomena that happen at much smaller scales. Strictly speaking, The upscaling procedure consists in determining the macroscopic parameter of interest, i.e. the diffusivity of water into the specified polymer mixture, starting from the information about the mean square displacement of the water molecules that are tracked. In particular, the water diffusion coefficient D_w of water molecules are obtained starting from their diffusion trajectories $r_i(t)$ and by applying the Einstein relation (3):

$$D_w = \frac{\langle |\vec{r_i}(t) - \vec{r_i}(0)|^2 \rangle}{6t} \tag{3}$$

where $\langle |\vec{r_i}(t) - \vec{r_i}(0)|^2 \rangle$ represents the root mean square displacement of the water molecules averaged over all possible time origins and t represents the time. In particular, starting from the macroscale analysis, the microscale model needs: amount of water in the mixture, polymer components weight fractions, and polymer weight average molecular mass. The subsequent downscaling allows to set up a new MD simulation starting from the information provided by the macroscopic model. By consequence this procedure determines the input data for the following MD simulation, and diffusion constant obtained is transferred to the following macroscale simulation. The development of a multiscale computational model requires the coupling of microscale and macroscale results by means of a time dependent iterative scheme of this form (4):

$$D_w(t_{n-1}) \Longrightarrow \rho_w(t_n,x), \rho_1(t_n,x), ..., \rho_i(t_n,x), ..., \rho_N(t_n,x) \tag{4}$$

The nanoscale models, which are investigated by means of MD simulation can be used used to obtain the diffusion coefficient for water; this parameter can then be incorporated into macroscale simulation to determine the polymer erosion rate, and this procedure can be iterated up till a given erosion rate is reached.

4 Diffusion of benzene within a PVA matrix

Lets go back to the first example, the benzene-PVA system, as starting point the set of bonded parameters for VA-VA beads and three sets of nonbonded parameters (VA-VA, benzene-benzene and VA-benzene) have to be calculated. For each set we run a full atomistic simulation of the two groups involved in the interaction and we calculated the free energy of interaction. The free energy calculations are performed using the Adaptative Biasing Force (ABF) framework Monticelli et al. (2008) as implemented in the NAMD code ((Darve and Pohorille, 2001; Darve et al., 2002)) and the all-atom CHARMM force field MacKerell et al. (1998). The ABF calculations provide the free energy profile as a function of the distance between the group of atoms. The atomistic MD simulations are carried out for a simulation time of 40 ns (time step of 1 fs) at a temperature of 300 K. The free energy of interaction, computed via the ABF framework, is monitored between the two groups of atoms as a function of their center-of-mass distance, in the range 0.2-2 nm using windows of 0.001 nm. The free energy profile of two bonded VA monomers is interpolated with a harmonic potential, giving an equilibrium distance of 0.276 nm and a force constant of 0.44036 $kcal/molnm^2$. On the other hand, from the interaction free energy profiles between nonbonded VA monomers, benzene molecules and VA-benzene molecules, we obtaine the three sets of r_0 (energy minimum) and ϵ (energy minimum depth), used to feed the Lennard-Jones 12-6 potential of the CG force field (see Table 1).

We can now generate two different molecular models: pure benzene and PVA with a small amount of benzene as penetrant molecule. All the systems are generated in the atomistic form using the Amorphous Cell construction tool of Materials Studio 4.4 (Accelrys, Inc.). The pure benzene

Nonbonded monomers	Nonbonded parameters	
	ε (kcal/mol)	r_0(Å)
VA – VA	1.306	4.115
VA - Benzene	0.957	5.51
Benzene - Benzene	0.9634	5.62

Figure 2. Nonbonded interactions for the three possible nonbonded monomers Gautieri et al. (2010b)

system contains 500 benzene molecules in a cubic periodic box whereas the second system contains 4 PVA chains (of 200 repeat units) and 12 benzene molecules in a cubic periodic box. The atomistic systems are then converted into CG systems using the Coarse Grainer tool of Materials Studio (Fig. 1). CG models are subjected to MD simulations using the Mesocite module and the MARTINI force field implemented in Materials Studio but modified including the parameters for bonded and nonbonded interactions. The CG systems are minimized for 1000 steps, then equilibrated for 1 ns (using a time step of 20 ps) at 300 K. Finally, production simulations are run for a simulated time of 10 ns and 200 ns (for pure benzene and for benzene in PVA matrix, respectively). During production simulations the diffusion trajectories of the benzene alone and benzene within the PVA matrix are used to obtain the diffusion coefficient D_i of a single permeant molecule i by applying the Einstein relation. The self diffusion coefficient of benzene which is obtained in by this approach is $2.273E-5cm^2s^{-1}$, very close to the experimental value of $2.203E-5cm^2s^{-1}$ Mills (2002). For the system consisted of 12 benzene molecules diffusing in a PVA matrix the obtained diffusion coefficient is $0.263E-5cm^2s^{-1}$ which is again in good agreement with the experimental value, that is $0.25E-5cm^2s^{-1}$ Pan et al. (2008). In conclusion, the obtained diffusion coefficients are in remarkable agreement with experimental data hence demonstrating that atomistic-informed CG simulations can be a valid approach to treat problems where the computational limits of classical MD simulations are too restrictive while, at the same time, strictly atomistic details are not mandatory. Thus, the multiscale approach extends the application range of computational modeling and provides a useful tool to investigate phenomena at the micro-scale which determine macroscopic physical properties and phenomena. Todays demand for materials with specific properties is rapidly overtaking the available technology, especially in the understanding of why many materials have their specific properties. When the molecular dimensions prevent from allowing full atom simulations the development of novel methods, which can consist also on a combination of atomistic and CG simulations in a multi-scale paradigm, are mandatory. Such techniques are nowadays applied for the study of biological problems, but could be applied also to study the diffusion of large non-biological molecules (which require a longer time to reach the normal diffusion regime), or the study of polymeric materials, for which representative domain is larger than a few cubic nanometers.

5 Degradation of a PLA matrix

Concerning the water transport properties of PLA matrices change during the degradation process one strategy consists in generating several atomistic models of PLA, with increasing level of water content (ϕ, modeling the swelling process) and with decreasing chain length (i, resulting from the hydrolysis process). In the present example, different bulk models with decreasing chain length account for the degraded state. The molecular models construction and subsequent MD simulations are performed using the Materials Studio 4.0 software (Accelrys, Inc) and the COMPASS force field Sun (1998). The polymeric chains with the desired length are generated starting from the repeating units using the Build Polymers tool of Materials Studio. Finally, we generate solvated amorphous bulk models containing PLA and water are generated using the Amorphous Cell tool of Materials Studio. 30 different models can be obtained by changing polymer length and/or water amount as summarized in Table 2.

	Quasi-dry	20%	40%	60%	80%
600	1/10	1/600	1/1600	1/3600	1/9600
300	2/10	2/600	2/1600	2/3600	2/9600
150	4/10	4/600	4/1600	4/3600	4/9600
75	8/10	8/600	8/1600	8/3600	8/9600
30	20/10	20/600	20/1600	20/3600	20/9600
1	600/10	600/600	600/1600	600/3600	600/9600

Figure 3. Molecular features of PVA-water systems in terms of number of molecules within each box

The polymeric chains are solvated in a periodic box with a total atom number varying from 6,000 atoms (for quasi-dry systems) to 35,000 atoms (for highly solvated systems). As an example two different systems are shown in Figure 4, the first is composed by two PLA chains of 300 monomers each and 600 water molecules; the second is characterized by four PVA chains of 150 monomers and 3600 water molecules.

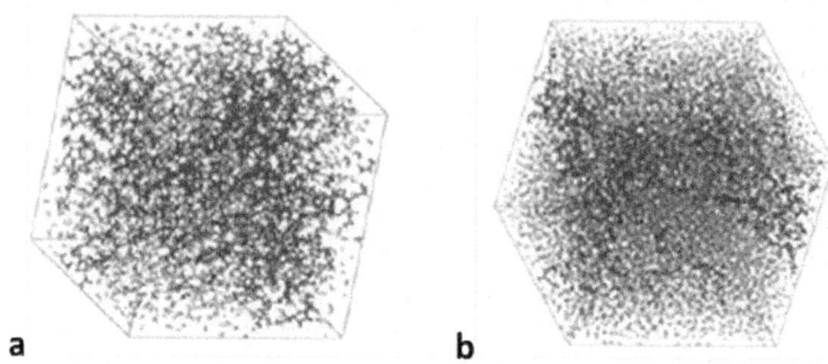

Figure 4. Atomistic models of two PVA-water systems.

In order to obtain water and polymer diffusivity NPT production simulation are performed for a simulated time of 7 ns. We then select an ensemble of M water molecules in the model and we compute their mean square displacement $MSD(t)$, denoting with:

$$\langle r(t)^2 \rangle = \frac{1}{M} \sum_{m=1}^{M} r_m(t)^2 \qquad (5)$$

the average displacement over all the particles D of water molecule is then calculated using Eq. 3. Water diffusivity in PLA spans two orders of magnitude (from 10^{-7} to $10^{-5} cm^2/s$) and it is mainly affected by the water content Gautieri et al. (2012). By informing the microscale model with nanoscale data, it is possible to obtain the evolution of the density of the bulk during the degradation process. In this regard the microscale model is capable to represent how the mean value of total mixture density evolves in the lumped state space. In particular, Figure5 show how starting from the density of the pure PLA (top, left) the microscale model correctly predict the reduction of this parameters reaching the value of the pure water (bottom, right).

Degradable materials offer tremendous potential for the development of implantable devices and systems for treating disease. The degradation in polymer is a transport-reaction mechanism, encompassing the transport of water in the polymer, the hydrolytic cleavage of the polymeric bonds, the transport of degradation products and their subsequent absorption in the

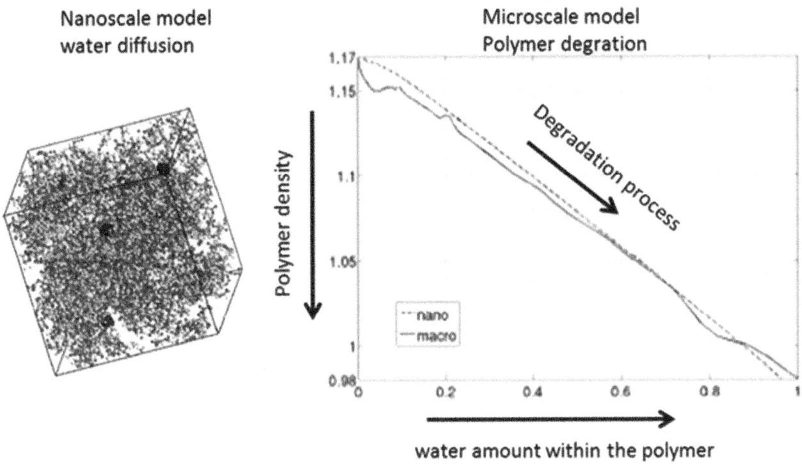

Figure 5. Atomistic bulk model of quasi-dry PLA matrix and predicted water/polymer mixture densities provided by the macro-scale model.

surrounding media.

6 Relevance of multiscale diffusion models in the biomedical field: design of drug delivery polymer

In the Biomedical field the need for new materials can be divided in two strategies: the design of new materials, and manipulation of the existing ones to obtaine a proper behavior. For their versatility polymers are one such class of materials and they found application for medical purpose in different biomedical fields such as drug delivery systems, scaffolds for tissue engineering, functionalized surface for medical devices, prosthesis and others. In all these applications one of the most challenging issue is to tailor te release of biochemical signals such as drugs, grow factors, differentiating factors etc. This partnership of polymer science and pharmaceutical science led to the innovation in the design and the development of novel drug delivery systems. The role of the polymers in such system is to deliver drugs to a particular target cell or tissue with a predefined/controlled kinetics in order to increase the effectiveness of the drug. The polymer core should have a well-defined properties such as biodegradability in order to protect the drugs from the physiological environment and to modulate the availability of the drug in the proper site of action. Due to their versatility different drug delivery systems have been developed such as responsive drug delivery systems, biodegrable drug delivery systems, diffusion controlled drug delivery systems. These polymeric delivery systems are differentiated depending on the controlling release mechanism of the drug from the polymer. Responsive drug delivery systems are a class of delivery systems where the rate of drug released can be controlled by its surroundings such as, Ph, temperature, ion concentration. For example copolymer of poly ethylene glycol and poly propylene glycol or copolymers of poly lactic acid and poly glycolic acid exhibit thermo-responsiveness and are currently used as thermogelling systems which are liquid at low temperature while gels at human body temperature, which eventually degrades and releases the drug molecule. Biodegradable drug delivery systems disassembly into biocompatible products when exposed to chemical (water) and/or biological (enzymes or microbial) factors, which, eroding the bulk polymeric material, leave exposed the drug to the cellular environment. Hence the drug molecule is released due to this process of erosion either in bulk or at the polymers surface. The degradation process is demonstrated by a molecular weight reduction due to the breakdown of the polymer chains. The biodegradability of these polymers can be easily manipulated by incorporating a class of labile groups such as ester, anhydride, carbonate, amide, in their backbone. Among this class we have polyanydrides, polyester, polyacrylic acids, polymethyl methacrylates, polyurethanes. The possibility to model for this class the biodegration timing would allow to study the effectiveness of a particular biogradable polymer for a given biomedical application. Diffusion controlled drug delivering systems involve the dispersion of therapeutic agent (drug) within a polymer shell. In this case the release of therapeutic agents by this system is driven by diffusion and the drug is realize either by passing through the pores or between polymer chains, and these are the phenomena that control the release rate. Here the dimensions of both the drug and the bulk turns the full atom resolution be impractical to obtain information about this mechanism. The solution presented here is to develop a CG method informed by full atomistic simulations to obtain realistic information about the diffusion of a large molecule within a polymer hence to be used as a model of the drug diffusion within a polymer. In conclusion, in order to advance from prototype status to a reliable human implant devices, device designers should prefer a preliminary modeling phase of the relevant phenomena instead of a combination of intuition and trial-and-error procedure that often fail due to two the lack of models able to describe the evolution or the behavior of the material. In Figure 4 a possible multiscale approach for the design of a biodegradable stent is sketched, for the different scales a specific output represents the input for the upper level in particular the diffusion coefficients of a drug obtained with full atom or CG model represents the imput for the microscale to obtain information about the drug release kinetics and eventually implementing this timing with a macroscale device it would be possible to model the behavior this device. In fact, if over the last decades experimental work has delivered important information to elucidate the basic phenomena involved in this process, an even more detailed understanding of the mechanism of degradation can be achieved only by combining this information with insight into the basic processes at an atomistic level.

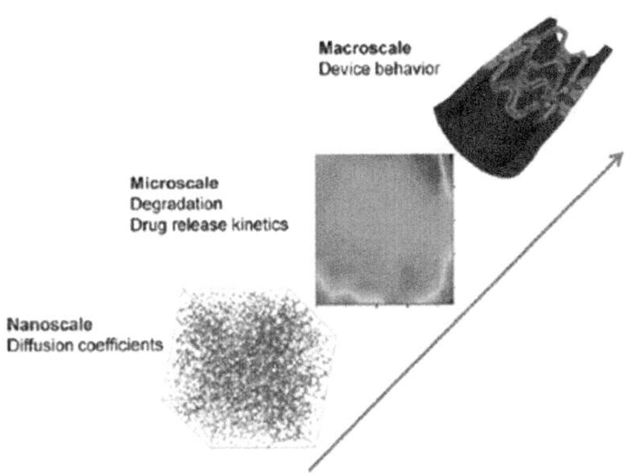

Figure 6. A multi-scale approach is envisioned for the design of biodegradable devices in which it is essential for precise control over the features of the polymeric matrix (e.g., drug release rates) during degradation process. Within this approach, atomistic simulations can provide the diffusivity coefficients of molecules (e.g., water, and drugs) for different states of the degrading matrix (Gautieri et al., 2010b, 2012). Using these data as an input, can then predict the degradation kinetics and thus how mechanical properties or drug release rates change with time. Finally, macroscale modeling would incorporate the information from the micro-scale to investigate the behavior of the final deviceSoares and Zunino (2010).

Bibliography

C. M. Agrawal and R. B. Ray. Biodegradable polymeric scaffolds for musculoskeletal tissue engineering. *Journal of Biomedical Materials Research*, 55(2):141–150, 2001.

E. Chiessi, F. Cavalieri, and G. Paradossi. Water and polymer dynamics in chemically cross-linked hydrogels of poly(vinyl alcohol): A molecular dynamics simulation study. *Journal Of Physical Chemistry B*, 111(11): 2820–2827, 2007.

Choon-Peng Chng and Lee-Wei Yang. Coarse-grained models reveal functional dynamics - ii. molecular dynamics simulation at the coarse-grained level - theories and biological applications. *Bioinform Biol Insights*, 2: 171–85, 2008.

E. Darve and A. Pohorille. Calculating free energies using average force. *Journal Of Chemical Physics*, 115(20):9169–9183, 2001.

E. Darve, M. A. Wilson, and A. Pohorille. Calculating free energies using a scaled-force molecular dynamics algorithm. *Molecular Simulation*, 28 (1-2):113–144, 2002.

A. Einstein. Von der molekulrkinetischen theorie der wrme gefordete bewegung von in ruhenden flssigkeiten suspendierten teilchen (the motion of elements suspended in static liquids as claimed in the molecular kinetic theory of heat). *Ann Phys*, 17:549560, 1905.

M. Entrialgo-Castano, A. Lendlein, and D. Hofmann. Molecular modeling investigations of dry and two water-swollen states of biodegradable polymers. *Advanced Engineering Materials*, 8(5):434–439, 2006.

A. Gautieri, S. Vesentini, and A. Redaelli. How to predict diffusion of medium-sized molecules in polymer matrices. from atomistic to coarse grain simulations. *J Mol Model*, 16(12):1845–51, 2010a. Gautieri, Alfonso Vesentini, Simone Redaelli, Alberto Research Support, Non-U.S. Gov't Germany Journal of molecular modeling J Mol Model. 2010 Dec;16(12):1845-51. Epub 2010 Mar 12.

A. Gautieri, S. Vesentini, and A. Redaelli. How to predict diffusion of medium-sized molecules in polymer matrices. from atomistic to coarse grain simulations. *J Mol Model*, 16(12):1845–51, 2010b. Gautieri, Alfonso Vesentini, Simone Redaelli, Alberto Research Support, Non-U.S. Gov't Germany Journal of molecular modeling J Mol Model. 2010 Dec;16(12):1845-51. Epub 2010 Mar 12.

A. Gautieri, A. Mezzanzanica, A. Motta, A. Redealli, and S. Vesentini. Atomistic modeling of water diffusion in hydrolytic biomaterials. *Journal of Molecular Modeling*, 18(4):1495–1502, 2012.

P. Gestoso and N. C. Karayiannis. Molecular simulation of the effect of temperature and architecture on polyethylene barrier properties. *Journal Of Physical Chemistry B*, 112(18):5646–5660, 2008.

D. Hofmann, M. Entrialgo-Castano, K. Kratz, and A. Lendlein. Knowledge-based approach towards hydrolytic degradation of polymer-based biomaterials. *Adv Mater*, 21(32-33):3237–45, 2009. Hofmann, Dieter Entrialgo-Castano, Maria Kratz, Karl Lendlein, Andreas Germany Advanced materials (Deerfield Beach, Fla.) Adv Mater. 2009 Sep 4;21(32-33):3237-45.

A. Lendlein and R. Langer. Biodegradable, elastic shape-memory polymers for potential biomedical applications. *Science*, 296(5573):1673–6, 2002. Lendlein, Andreas Langer, Robert Research Support, Non-U.S. Gov't United States Science (New York, N.Y.) Science. 2002 May 31;296(5573):1673-6. Epub 2002 Apr 25.

A. D. MacKerell, D. Bashford, M. Bellott, R. L. Dunbrack, J. D. Evanseck, M. J. Field, S. Fischer, J. Gao, H. Guo, S. Ha, D. Joseph-McCarthy, L. Kuchnir, K. Kuczera, F. T. K. Lau, C. Mattos, S. Michnick, T. Ngo, D. T. Nguyen, B. Prodhom, W. E. Reiher, B. Roux, M. Schlenkrich, J. C. Smith, R. Stote, J. Straub, M. Watanabe, J. Wiorkiewicz-Kuczera, D. Yin, and M. Karplus. All-atom empirical potential for molecular modeling and dynamics studies of proteins. *Journal Of Physical Chemistry B*, 102(18):3586–3616, 1998.

S. J. Marrink, A. H. de Vries, and A. E. Mark. Coarse grained model for semiquantitative lipid simulations. *Journal Of Physical Chemistry B*, 108(2):750–760, 2004.

R. Mills. Search for isotope effects in the self-diffusion of benzene and cyclohexane at 25.deg. *The Journal of Physical Chemistry*, 79(8):852–853, 2002. doi: 10.1021/j100575a016.

L. Monticelli, S. K. Kandasamy, X. Periole, R. G. Larson, D. P. Tieleman, and S. J. Marrink. The martini coarse-grained force field: Extension to proteins. *Journal of Chemical Theory and Computation*, 4(5):819–834, 2008.

F. S. Pan, F. B. Peng, L. Y. Lu, J. T. Wang, and Z. Y. Jiang. Molecular simulation on penetrants diffusion at the interface region of organic-inorganic hybrid membranes. *Chemical Engineering Science*, 63(4):1072–1080, 2008.

D. Pavel and R. Shanks. Molecular dynamics simulation of diffusion of o-2 and co2 in blends of amorphous poly(ethylene terephthalate) and related polyesters. *Polymer*, 46(16):6135–6147, 2005.

D. Pavel and R. Shanks. Molecular dynamics simulation of diffusion of o-2 and co2 in amorphous poly(ethylene terephthalate) and related aromatic polyesters. *Polymer*, 44(21):6713–6724, 2003.

B. Prathab and T. M. Aminabhavi. Molecular modeling study on surface, thermal, mechanical and gas diffusion properties of chitosan. *Journal of Polymer Science Part B-Polymer Physics*, 45(11):1260–1270, 2007.

J. S. Soares and P. Zunino. A mixture model for water uptake, degradation, erosion and drug release from polydisperse polymeric networks. *Biomaterials*, 31(11):3032–42, 2010. Soares, Joao S Zunino, Paolo Research Support, Non-U.S. Gov't England Biomaterials Biomaterials. 2010 Apr;31(11):3032-42. Epub 2010 Feb 2.

H. Sun. Compass: An ab initio force-field optimized for condensed-phase applications - overview with details on alkane and benzene compounds. *Journal Of Physical Chemistry B*, 102(38):7338–7364, 1998.

J. A. Tamada and R. Langer. Erosion kinetics of hydrolytically degradable polymers. *Proc Natl Acad Sci U S A*, 90(2):552–6, 1993. Tamada, J A Langer, R CA 52857/CA/NCI NIH HHS/United States GM44884/GM/NIGMS NIH HHS/United States Comparative Study Research Support, U.S. Gov't, P.H.S. United states Proceedings of the National Academy of Sciences of the United States of America Proc Natl Acad Sci U S A. 1993 Jan 15;90(2):552-6.

V. Tozzini. Coarse-grained models for proteins. *Current Opinion In Structural Biology*, 15(2):144–150, 2005.

Advances in Experimental Cell Biology and Cell-Material Interactions

Claire M. Cobley[†], Seraphine V. Wegner[†], Martin Streichfuss[†] and Joachim P. Spatz[†]

[†] Department of Biophysical Chemistry, University of Heidelberg, Heidelberg, Germany
and
Department of New Materials and Biosystems, Max-Planck-Institute for Intelligent Systems, Stuttgart, Germany

Abstract Recent advances in the physical sciences and engineering have made it possible to measure and manipulate the mechanical and binding properties of cells in new ways. In this chapter, we will introduce this field by discussing two different experimental approaches at the interface of biology and engineering: 1) ways to measure the different types of forces generated by the actin cytoskeleton and 2) how we can probe the interactions between cells and their environment using nanostructured surfaces.

1 Introduction

A single cell could be considered the fundamental building block of life. Yet within this single unit there exist innumerable mechanisms for the cell to sense and interact with its environment. These processes involve many different types of signaling mechanisms, from molecular cues to physical forces (Alberts et al., 2008).

The primary system responsible for regulating cellular mechanics is the cytoskeleton. Far from just acting as a static scaffold to maintain the structural integrity of the cell, the cytoskeleton is a remarkably adaptive and highly dynamic structure that provides the forces necessary for the cell to move, reproduce, and react to external stimuli (Janmey, 1998). For example, it is the main determinant of the cell's stiffness and orientation and plays a key role in cellular processes such as migration and division. The cytoskeleton is composed of three structure-giving proteins, namely actin, keratin and tubulin. All three exist in monomeric and polymerized forms and can be organized into higher order structures such as bundles and networks by specific binding and crosslinking proteins. In this way simple components can become a complex, flexible system (Alberts et al., 2008).

Over the years, experimentalists have developed a variety of tools to probe the mechanical properties of cells on different scales. For example, experimental setups have been developed to investigate a cell's response to membrane stretching (by pulling on the substrate) or to probe the cell's stiffness in different regions (through indentation with an atomic force microscopy tip) (Jungbauer et al., 2008; Huang et al., 2004). On a much smaller scale, it is also possible to isolate the individual components of the cytoskeleton, and study their fundamental mechanical properties in controlled environments outside of the cell (Streichfuss et al., 2011).

Herein, we will illustrate through case studies some of the techniques used to understand both cellular forces and cell-material interactions. We will start with one of the fundamental building blocks of the cytoskeleton, actin filaments, and build up to more complex, whole-cell interactions. We hope that in this way we can provide a flavor of this rich field of research.

2 Forces in the Actin Cytoskeleton

As it is the key player in cell motility and integrity, the actin cytoskeleton has been intensively studied on levels ranging from a single filament to groups of filaments (e.g. bundles, networks and gels) to the entire cytoskeleton (the whole cell). The first part of this chapter focuses on measurements of the actin-based force generation driving cell motility and a biomimetic model of the actin cortex, a quasi-two-dimensional network beneath the plasma membrane of eukaryotic cells.

Actin exists both in monomeric and polymerized forms. Actin filaments are 7-9 nm thick and are built up from the monomeric building block G-actin, a globular protein consisting of two domains. When G-actin concentrations are over 0.1 μM or the salt concentration is high, the G-actin starts to polymerize at the plus (aka barbed) end of the polar polymer, forming filamentous actin. This polymerization is not only responsible for the presence of filamentous actin in the cell, but is also the driving mechanism behind membrane protrusions (i.e. lamellipodia and filopodia) and bacterial locomotion. Unlike molecular motors, which consume chemically stored energy to generate forces, this so-called "brownian ratchet" or "polymerization motor" relies exclusively on Brownian motion (Peskin et al., 1993). Either the filament itself or the obstacle (e.g. the cell membrane) fluctuates according to its thermal motion, thereby generating a gap sufficiently large for another G-actin to be added to the growing filament in front of the obstacle (Figure 1). As the filament grows monomer by monomer, it exerts a force against the obstacle.

The forces generated by this mechanism have been studied both exper-

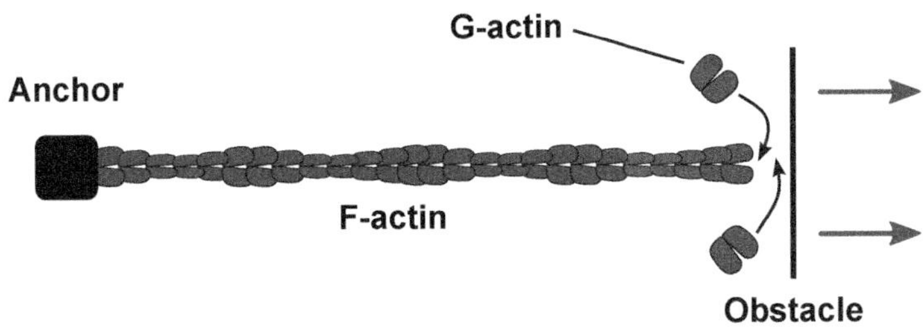

Figure 1. Schematic of the polymerization motor. The slow-growing (minus/pointed) end of the filament is anchored (black box), whereas at the fast growing (plus/barbed) end of the filament, addition of G-actin is still possible. An obstacle (black bar) limits the growth of the filament as soon as the gap between the growing filament and the obstacle becomes smaller than the size of a globular subunit. As the obstacle undergoes Brownian motion, the spacing becomes large enough for another subunit to be added. Therefore, the growing filament exerts a force (red arrows) along its axis pushing the obstacle.

imentally and theoretically. In addition to membrane protrusions in eukaryotic cells, pathogenic bacteria such as Listeria are believed to use the polymerization motor to move within a host cell. The pathogen presents ActA, a protein promoting actin polymerization, on its surface, inducing fast filament growth and movement. A minimal system of an ActA coated microsphere in an actin-rich cytoplasmic extract is sufficient to propel a 0.5 μm microsphere against the frictional force in medium with a speed of up to 0.2 μm/second (Cameron et al., 1999). A comet tail of several actin filaments behind the microspheres or bacteria is visible with both fluorescence imaging and electron micrographs (Brieher et al., 2004; Cameron et al., 1999). The question what force a single growing filament can contribute to this amazing propulsion can be calculated with

$$f = \frac{k_B T}{\delta} \ln\left(\frac{c}{c_c}\right) \tag{1}$$

where k_B is the Boltzmann factor, T the temperature, δ the gap size needed for an actin subunit to be added, c the concentration of G-actin and c_c the critical concentration (Peskin et al., 1993). It is also possible to measure the force of a single growing filament directly using optical tweezers (Footer et al., 2007). Optical tweezers are a versatile tool for force measurements in biology because they can sense forces from less than a piconewton up to hundreds of piconewtons with high spatial resolution. In addition, they can be operated in biologically relevant medium to avoid damaging the specimen.

In one set of experiments, a bundle of actin filaments was attached to a microsphere trapped by optical tweezers and the force exerted by the polymerization motor against a rigid barrier was directly measured (Footer et al., 2007). The experimentally determined value of 0.76±0.22 pN corresponds very well with the calculated value of 0.8 pN for the maximum force (eq. 1).

It is also important that the growing semi-flexible filament can bear the force it exerts on the obstacle. In order to estimate the maximum force the filament can take before it buckles, one can use the Euler-Bernouille beam theory. The critical force at which a semi-flexible filament buckles is given by the critical Euler buckling force

$$f_{\text{crit}} = \frac{\pi^2 \cdot EI}{(KL)^2} = \frac{\pi^2 \cdot l_p k_B T}{(KL)^2} \tag{2}$$

where E is the elastic modulus, I the second moment of area, K a geometrical factor and L the length of the polymer. K depends on the condition of the ends of the polymer: In the case in Figure 1, one end is pinned to the anchor and the other is free to move laterally corresponding to $K = 2$. Replacing EI with the more common expression for semi-flexible polymers $l_p k_B T$, where l_p is the persistence length ($\approx 15\ \mu m$ for filamentous actin), gives a critical load of 0.15 pN for a 1 μm long actin filament. Therefore, it is clear that such a filament could not bear the load of 0.8 pN. However, the critical load of a bundle of filaments scales with N^α, where N is the number of filaments and α depends on the interconnection between the filaments. For weakly attached filaments, the force scales linearly with N ($\alpha = 1$), but for bundles bound by the actin binding protein (ABP) fascin, it scales quadratically ($\alpha = 2$, Mogilner and Rubinstein (2005)). This might be the reason for the fact that 10-30 parallel actin filaments bundled by fascin were found in filopodial protrusions (Figure 2) (Svitkina et al., 2003).

The next important issue to consider is how much force is needed to deform the cell membrane. Based on a Helfrich model, one can calculate the force of a cell membrane forming a cylindrical protrusion as

$$F = \frac{\kappa\pi}{R} + 2\pi R\sigma \tag{3}$$

where R is the radius of the protrusion, κ the membrane bending stiffness and σ the surface tension (Pronk et al., 2008; Nambiar et al., 2010). Using

$R = 200$ nm, $\kappa = 270$ pN nm and $\sigma = 0.003\ \frac{\text{pN}}{\text{nm}}$, this equation gives a force of about 8 pN (Nambiar et al., 2010). This corresponds very well to the force of 10-30 filaments growing in a filopodium.

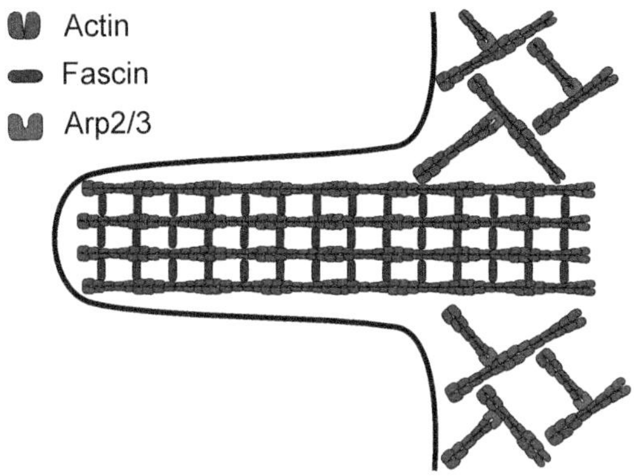

Figure 2. Schematic drawing of a filopodium. The cell membrane (black) is pushed by a growing bundle of several actin filaments (red). The filopodial actin filaments originate from a branched network of actin in the lamellipodium, which is organized by the Arp2/3 complex. Fascin (blue) bundles the filaments tightly, thereby greatly enhancing the mechanical stability of the bundle.

Another mechanism possibly driving membrane protrusions and bacterial locomotion has been proposed based on attractive interactions between two semi-flexible actin filaments in a splayed configuration and in contact with the obstacle being pushed (Figure 3) (Kühne et al., 2009; Streichfuss et al., 2011).

At the barrier the two filaments are splayed, whereas at the anchor they are bundled by an attractive interaction. This attractive interaction can be induced by counterion condensation along the negatively charged actin filament, ABPs, or the crowded cellular environment. In our group, we have used holographic optical tweezers (HOTs) to study the forces generated by the bundling of two individual actin filaments in such a geometry. Holographic optical tweezers use a diffractive optical element to alter the phase of the incoming laser beam in such a way that multiple traps are generated in the focal plane of the objective. These traps can be switched on and off, moved with high precision, tuned in intensity, and used as very sensitive force sensors. In combination with a microfluidic device, we can control the physical and chemical environment very precisely in order to study the influence of ionic strength on the force generation. The experiment started by

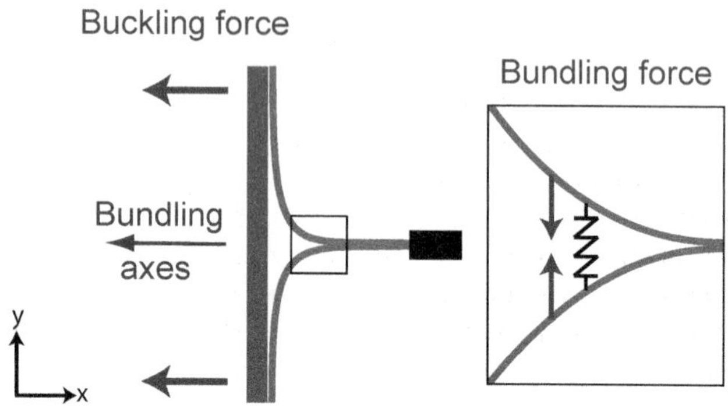

Figure 3. Schematic drawing of two actin filaments pushing an obstacle. Two actin filaments are anchored (black box) on one end. The opposite ends are splayed and in contact with the barrier to be pushed. The attractive interactions between the two filaments cause a force bundling them together (bundling force in the inset). Due to the semi-flexible nature of filamentous actin the bending is thereby increased and the buckled filaments exert a force along the bundling axes against the barrier (buckling force).

trapping three microspheres with HOTs, attaching two filaments to them in the configuration shown above, and then inducing the bundling with divalent ions (Figure 4). The forces driving bundle formation were measured by repeatedly changing the potential of the optical traps and recording the corresponding displacement of the microspheres (Figure 4B). After calibration of the optical traps, we could calculate the force exerted by the attractive interaction between the two filaments.

We also used a complementary approach to determine the forces during dynamic bundle formation. Having prepared the steady state condition described above, we switched off the optical traps for microspheres 1 and 2 (Figure 4A). Due to the attractive interaction, the filaments bundled and consequently pulled the microspheres against the viscous drag. Knowing the velocity of the dragged microsphere, we used the Stokes equation to calculate the generated force. These two complementary methods demonstrated independently that the forces generated depend on the concentration of divalent ions and can go up to 0.2 pN at 100 mM (Figure 5).

In addition to these fundamental measurements of the interactions between two filaments, we conducted experiments using multiple actin filaments on a scaffold of seven microspheres that resembles the actin cortex, a quasi two-dimensional network of actin beneath the plasma membrane

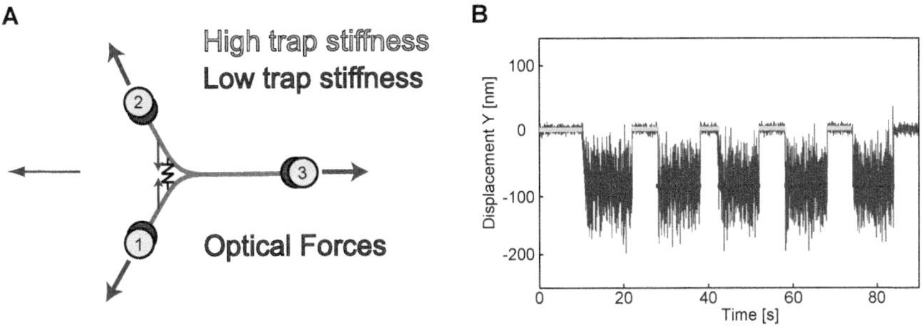

Figure 4. Schematic drawing of the experiment. First, we trap three microspheres with HOTs and attach two single actin filaments such that one is connected to microsphere 1 and 3, the other to microsphere 2 and 3. Subsequently, the bundling is induced by moving the filaments with the microspheres to a buffer with divalent ions, which results in bundling of the filaments along the bundling axes (blue) as far as the optical traps allow for. In this steady state, the optical forces (red) balance the attractive interaction. Next, the laser intensity and therefore the potential of the optical traps is alternately lowered and raised from $\approx$ 10 pN/μm (yellow) to $\approx$ 1 pN/μm (red), with simultaneous monitoring of the displacements of the microspheres. An example of the displacement of one microsphere is given in B.

(Uhrig et al., 2009). These seven microspheres were held by HOTs in a hexagonal array and subsequently filamentous actin was attached. Next, a buffer with 100 mM Mg^{2+} was introduced into an exchange chamber, allowing for the diffusion of the divalent ions into the experimental chamber. As soon as the ions reached the network, the filaments started to bundle and the networks contracted (Figure 6). Due to this contraction, the optically trapped microspheres were displaced, and the exerted force could be determined after trap calibration. These contractile forces increased as more and more ions diffused into the network, reaching up to 3 pN. This implies that more than 10 filaments were pulling on a microsphere, as previous experiments showed 0.2 pN per filament pair (Figure 6D). For the central microsphere the forces do not balance entirely as the filaments are not homogeneously distributed over the hexagon.

In summary, the measurements of force generation between two actin filaments during bundle formation revealed forces of up to 0.2 pN with 100 mM divalent ions, which is less than the force exerted by a single grow-

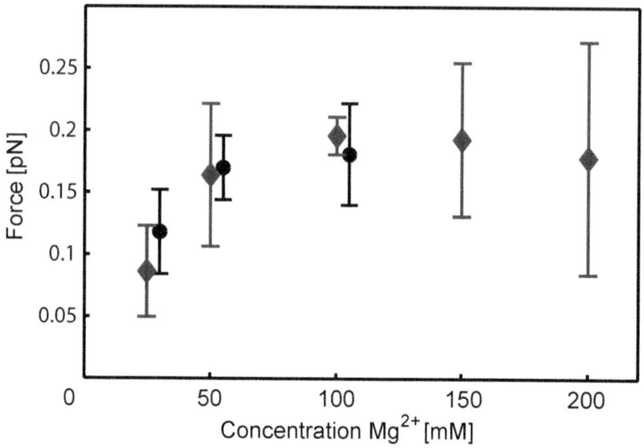

Figure 5. Forces generated by the bundling of two individual actin filaments, as measured by optical tweezers. The forces were measured in the steady state (red) and during dynamic bundle formation (black) using different concentrations of Mg^{2+} to induce counterion condensation along the negatively charged actin filaments. The forces are given by the mean ± standard deviation.

ing filament. However, while scaling of the buckling force with increasing number of filaments is expected and has been treated in theory, it has not yet been investigated experimentally (Kühne et al., 2009). A related experimental approach with 2D networks of actin filaments on a scaffold of 7 optically trapped particles in a hexagonal arrangement supports this idea, as the forces increased up to 3 pN for each of the trapped microspheres due to the large number of attached actin filaments (Uhrig et al., 2009).

3 Cell-Substrate Interactions

The actin network and the rest of the cytoskeleton are linked to the extracellular environment through large assemblies of proteins known as focal adhesions (Zamir and Geiger, 2001). Focal adhesions transmit information about the substrate to the actin network through transmembrane proteins called integrins. Integrins have both an extracellular component, which associates with proteins in the extracellular matrix surrounding the cell, and an intracellular component, which associates with the actin cytoskeleton

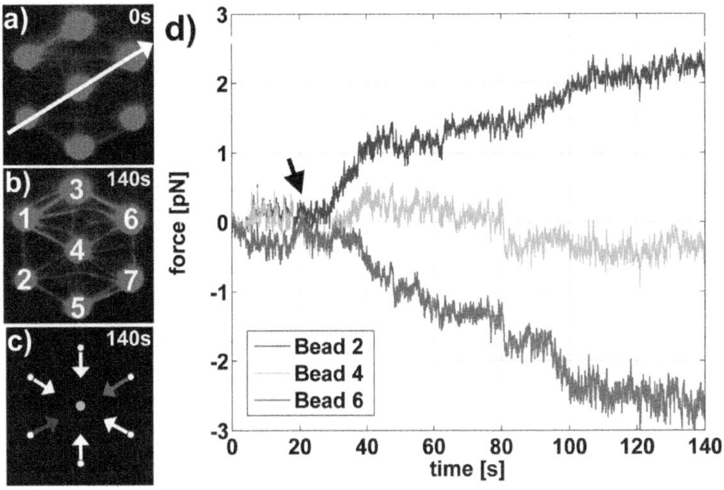

Figure 6. Experimental approach addressing the contractile forces in a quasi 2D actin network. A) Actin network with loosely hanging actin filaments on a hexagonal scaffold of microspheres optically trapped by HOTs. B) After diffusion of Mg^{2+} to the network the filaments start to bundle and contract the network. C) Schematic representation of the scaffold with the three microspheres colored red, yellow, and green, whose forces are shown in D. D) Forces exerted on microspheres 2, 4, and 6 by the contracting actin network during the bundling process induced by diffusion of Mg^{2+} ions to the network.

through focal adhesions. These linkages allow the cell to sense and respond to its mechanical environment (Giancotti and Ruoslahti, 1999). The information the cell receives about the substrate is transmitted throughout the cell via signaling pathways that can activate specific proteins or modify gene transcription. These signals (combined with others) influence whether the cell will grow, migrate, divide, differentiate, or even begin programmed cell death (Geiger et al., 2001).

The importance of this signaling is evident in the sensitivity of cells to the properties of the surface on which they are growing. Most cells from solid tissue must attach to a substrate before they are able to grow and divide. While there is a lot of variation between cell types, it can generally be said that cells attach more strongly to substrates with a similar stiffness to the tissue they came from (e.g. brain tissue 0.1-1 kPa and muscle 8-17 kPa) and in the presence of extracellular matrix proteins (Engler et al.,

2006; Ruoslahti and Pierschbacher, 1987). This has been nicely illustrated by plating cells on surfaces which have been micropatterned with specific shapes of adhesive molecules; the cells only adhered in the pre-coated regions and stress fibers (bundles of actin filaments) formed along the edges of the pattern (Théry et al., 2006).

However, many questions still remain about the precise nature of the cell-substrate interaction, particularly about the spatial arrangement of adhesion sites. It is difficult to perform studies on this topic because cells are constantly changing their environment by excreting extracellular matrix proteins and because many of the processes happen below the resolution limit of typical microscopy setups. In particular, researchers are interested in understanding better the role that integrin clustering plays in cell signaling. It has been shown that different signaling pathways are activated when only one integrin or a cluster of integrins is activated (Miyamoto et al., 1995), but the specifics of the receptor arrangement in these clusters is still not completely understood.

Towards this end, our group developed nanopatterned substrates that can present integrin-binding ligands with controlled densities and spacings (Girard et al., 2007; Cavalcanti-Adam et al., 2008). These surfaces are created through a technique known as block copolymer micelle lithography (BCML) (Lohmüller et al., 2011; Arnold et al., 2004). In the first step, gold salt-loaded micelles are created using block copolymers of polystyrene (PS) and poly-2-vinylpyridine (P2VP). Block copolymers are polymers with clearly defined segments comprising different monomer building blocks. In the block copolymers used here, the different components have dramatically different chemical properties: PS is nonpolar, while P2VP is polar. This heterogeneity within the copolymer leads to the formation of inverse micelles when the copolymer is dissolved in nonpolar toluene; the P2VP ends cluster together, creating a sphere with a polar core and a nonpolar shell. When a gold salt is added to this solution, the gold ions preferentially cluster in the polar cores. In the next step, a clean glass surface is dipped into the solution, generating a monolayer of micelles arranged in a hexagonal array as the glass slide is slowly removed from the micellar solution. It is important to precisely control the dipping speed in order to ensure well-ordered arrays.

This surface is then treated with an oxygen plasma, which reduces the gold salt to generate solid gold nanoparticles between 1 and 12 nm in diameter, depending on the amount of gold salt loaded. The plasma treatment also removes the excess solvent and the copolymer from both the gold nanoparticles and the glass substrate. This technique is capable of producing gold nanoparticle arrays with well-defined inter-particle distances and a high degree of order, as shown in Figure 3. The inter-particle distance can be tuned between 30 and 250 nm by tuning the copolymer composition, concentrations, and the dipping speed (Spatz et al., 2000; Lohmüller et al., 2011).

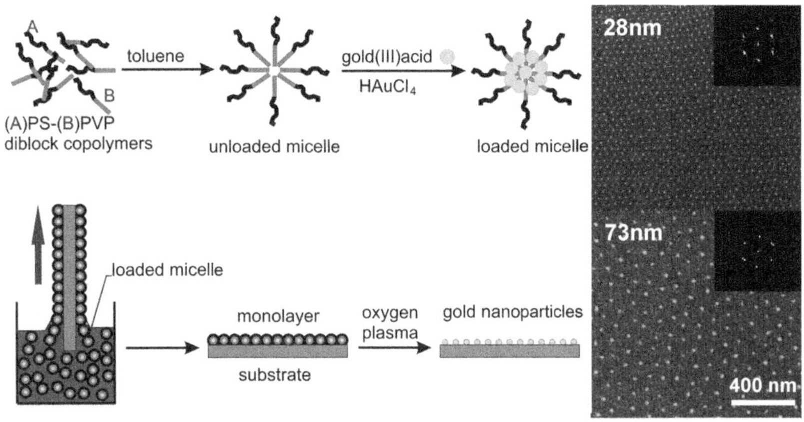

Figure 7. Block copolymer micelle lithography. Clean surfaces are coated with a monolayer of gold ion-loaded micelles by dipping the substrate under precisely controlled conditions. After treatment with an oxygen plasma, the surfaces are patterned with a precise array of gold nanoparticles.

We chose to create the nanoparticle arrays with gold due to the high biocompatibility and well-known surface chemistry of this metal. Thiol group-containing compounds (R-SH) bind readily to gold and have been used to attach a wide variety of biologically active components to gold surfaces and nanoparticles. For integrin-related experiments, we conjugate the nanoparticles with a cyclic-RGD peptide containing a thiol group at the end of a short alkane spacer. The RGD amino acid sequence (Arginine-Glycine-Aspartic acid) is commonly found in extracellular matrix proteins and is known to bind to integrin receptors (Ruoslahti and Pierschbacher, 1987). Cells typically adhere readily on surfaces displaying these amino acid sequences. By placing these peptides only on the surface of the nanoparticles, we can study how the spacing of RGD ligands affects the cell's response. As the cross-section of an integrin is 10-12 nm, when small gold nanoparticles are used (e.g. 5-8 nm), it can be assumed that only one integrin can bind to each nanoparticle (Figure 8A) (Arnold et al., 2004). Therefore, the spacing of the gold nanoparticles on the surface controls the closest distance two integrins can obtain, consequently controlling their clustering density. In order to ensure that the ECM proteins secreted by the cell do not adsorb onto the glass between the nanoparticles and lead to non-specific interactions, these regions are passivated with covalently-bound poly(ethylene glycol) (PEG), which repels proteins and provides a neutral background (Bluemmel et al., 2007). With this setup, the cells can only interact with the surface through the nanopatterned RGD peptides. The specific interaction between the cell and the RGD peptides on the gold nanoparticles was observed in SEM images of cells fixed after they had adhered to nanopatterned surfaces (Figure 8B). Protrusions extend from the edge of the cell to the nanoparticles (visible due to dehydration during the fixing process), but not to the surface in between.

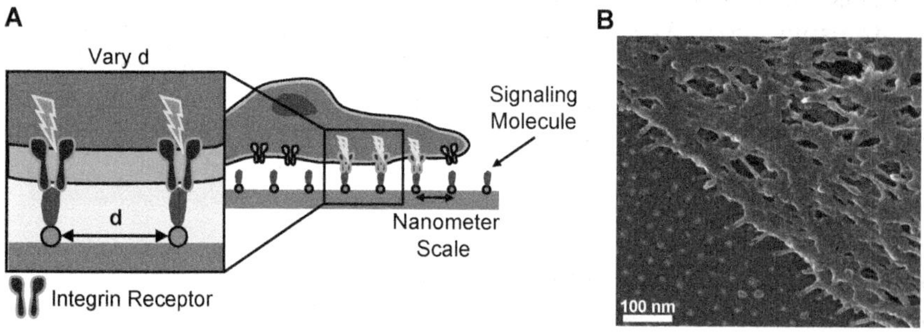

Figure 8. A) Integrins are 10-12 nm in size and therefore only one integrin can bind to one gold nanoparticle, which has a size of 5-8 nm. The interparticle distance on the surface also controls the closest spacing two integrins can have when cells adhere. B) The interaction of the cell with the gold nanoparticles and not the PEG-passivated background can be seen under the SEM.

The importance of integrin clustering for cell adhesion was studied by plating rat embryonic fibroblast (REF) cells on RGD-functionalized nanopatterned surfaces with different gold nanoparticle spacings. When the distance between the gold nanoparticles was less than 60 nm, a large number of cells could adhere to the surface. However, when the spacing was larger than 70 nm, the number of cells that could adhere was greatly reduced (Figure 9A). Differences could also be seen on the single cell level; cells on surfaces with gold nanoparticle spacings under 60 nm had a larger spreading area and were more circular in shape. Furthermore, when the cells were stained for the integrin β_3 (blue), the focal adhesion protein vinculin (red) and the actin cytoskeleton (green), cells on surfaces with spacings under 60 nm showed colocalization of vinculin and integrin β_3, indicating the formation of mature focal adhesions. The formation of the focal adhesions resulted in the establishment of distinct stress fibers in the actin cytoskeleton, which were also only observed on the surfaces with particle spacings below 60 nm (Figure 9C-D) (Cavalcanti-Adam et al., 2006). The cells on the surfaces with a larger spacing had contact with only a few spots on the surface, were more triangular in shape, had a smaller spreading area and didn't form mature focal adhesions (i.e. vinculin and integrin β_3 stains were diffuse and did not colocalize) (Figure 9C-D). From this data we concluded that integrins need to cluster closer together than 60 nm for mature focal adhesions to form (Figure 9B) (Arnold et al., 2004).

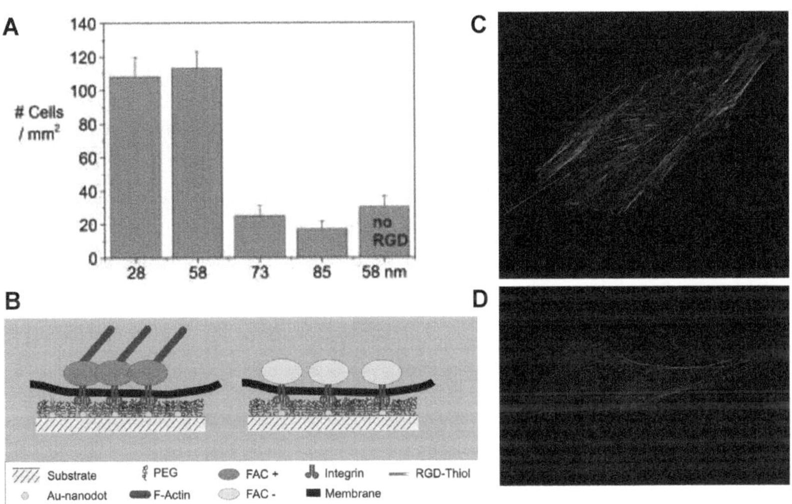

Figure 9. Adhesion of cells on gold nanopatterns with different particle spacings. A) Cells adhere well on surfaces with inter-particle distances lower than 60 nm but not on surfaces with larger particle spacings. B) Mature focal adhesions can form and connect to the actin skeleton on surfaces where the integrins can cluster closer than 58 nm. Focal adhesions cannot form when the spacing of integrins is larger than 73 nm. Overlayed images of cells on surfaces with 58 nm (B) and 109 nm (C) inter-particle spacing stained for integrin β_3 (blue), vinculin (red), actin (green).

Cells can sense not only the overall spacing but also gradients of the adhesion ligand RGD (Hirschfeld-Warneken et al., 2008). The spacing of gold nanoparticles can be varied on a glass surface by changing the speed the glass is removed from the gold micellar solution. By pulling with a constant acceleration, gradients of gold nanoparticles with a controlled steepness can be formed. Cells were plated on such surfaces with 15 nm/mm gradients (i.e. the inter-particle distance changes 15 nm per mm on the slide), resulting in a spacing that increased from 60 nm up to 100 nm. On these surfaces, cells were able to sense the gradient and moved towards the closer-spaced part of the surface, thereby adopting a polarized shape. Based on the size of a cell and the steepness of the gradient, this means that the cell can sense a difference of less than 1 nm in the ligand spacing between the opposing ends of the cell. Interestingly, the cell adhesion behavior on ordered vs. randomly patterned gold nanoparticle surfaces was strikingly different (Huang et al., 2009). To study this phenomenon in detail, four surfaces were prepared with gold nanoparticles with high density ordered, high density random, low density ordered and low density random patterns (Figure 10A). Cells showed similar adhesion to surfaces with a high ligand density whether the particles were ordered or random. However, at low particle densities, cells couldn't adhere to the ordered substrates, while the cells were able to adhere if the nanoparticle arrangement was random (Figure 10B). Thus, even if gold nanoparticle, and therefore adhesion molecule densities, are low on a global level, the cell can adhere if localized clusters can be formed.

As becomes clear from the comparison between ordered and random patterned surfaces, not the global but the local clustering of integrins is critical for the formation of an adhesion cluster. Therefore, the question becomes how many integrins are required in a cluster for one stable focal adhesion to form (Arnold et al., 2009). To answer this question, we have to control the spacing of the gold nanoparticles not only on the nanometer length scale but also on the micrometer length scale. To produce such surfaces, a combination of the above described BCML technique and photo/e-beam lithography was used (Figure 11A). In short, the gold nanostructured glass surfaces prepared by BCML were coated with a photo-sensitive resist which can be etched away by a light/e-beam into μm-sized patterns. Nanoparticles that are no longer protected by the resist can selectively be removed. Finally, the rest of the resist is also removed, yielding nanopatterned microstructures on glass.

Nanopatterns with 58 nm gold nanoparticle spacings were divided into 150 x 150 μm patches with square micropatterns (3000 nm, 1000 nm, 500 nm, 250 nm and 100 nm in side length) that were separated by their respective side length (Figure 11B). Therefore, the number of gold nanoparticles in each square was controlled by the size of the square micropattern, and the number of gold particles per square was 3000, 300, 80, 30 or 6.

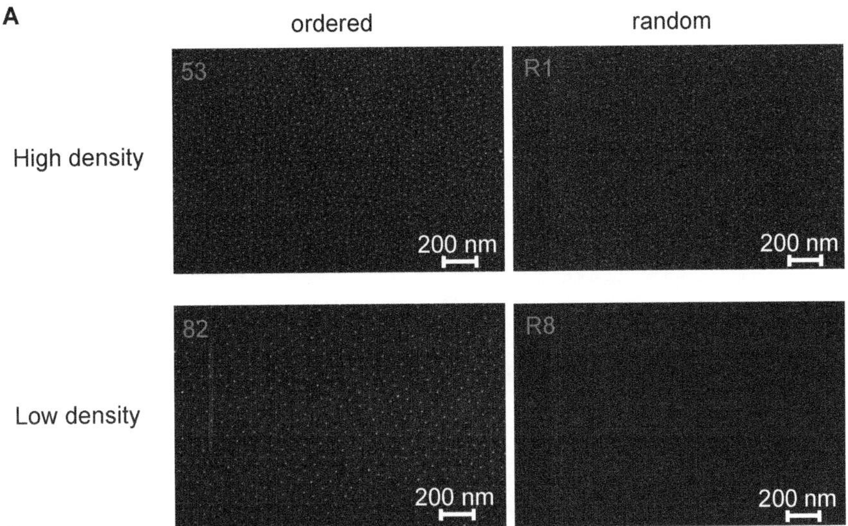

B

Pattern	Au nanoparticle density (μm^{-2})	Average Interdot Spacing±Standard Deviation (nm)	**Cell Adhesion**
R1: Random High density (1)	408 ± 32	55 ± 14	**YES**
R8: Random Low density (2)	151 ± 39	83 ± 33	**YES**
53: Ordered High density (3)	416 ± 20	53 ± 8	**YES**
82: Ordered Low density (4)	171 ± 16	82 ± 12	**NO**

Figure 10. Cell adhesion on ordered vs. random gold nanoparticle patterns. A) SEM images of the prepared surfaces. B) Characterization of the ordered and randomly patterned surfaces.

Despite the dramatically reduced global nanoparticle density, the cells were able to adhere to these micro-nanopatterns. However, differences were observed. In the case of the larger patches with 3000 and 300 gold nanoparticles per patch, formation of classical focal adhesions on the patches was observed. (Figure 11C) On the other hand, while the cells were still able to adhere to the smaller patches, with 80, 30 or 6 nanoparticles per patch, elongated adhesions bridging several patches were observed (Figure 11D). When the number of gold nanoparticles in a square was lower than 6 per

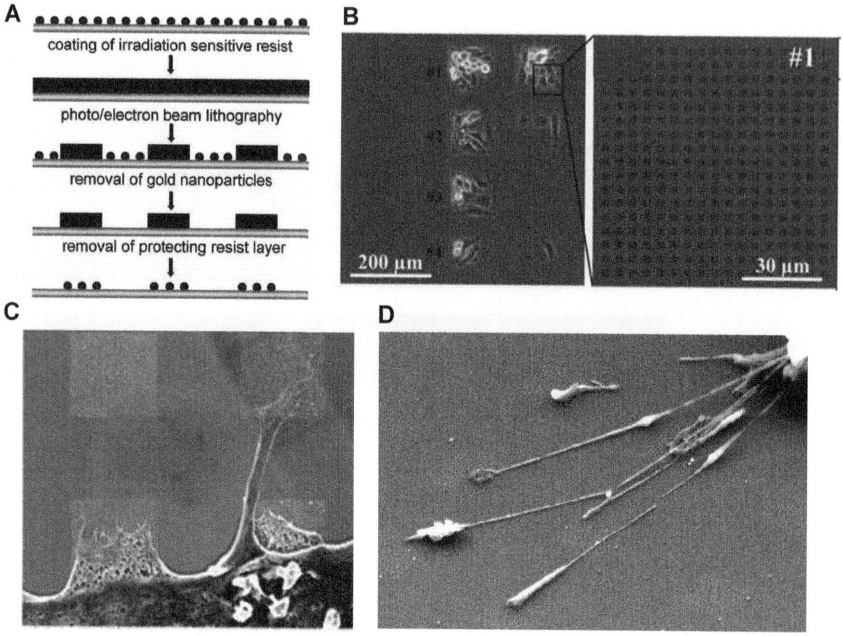

Figure 11. Cell adhesion on micro-nanopatterned gold nanoparticle surfaces. A) Preparation of micropatterns. B) Micropattern on the surfaces. C) Cells on nanostructured 3000 x 3000 nm micropatterns. D) Cells on nanostructured 250 x 250 nm micropatterns.

patch, the cells could not form stable focal adhesions, which shows that 6 is the minimum number of integrins that have to cluster in order to form a stable focal adhesion. Together these experiments help shed a light on the complex interactions between cells and their substrates. Through the use of carefully designed nano- and micro-structured materials, we were able to show that two integrins have to be closer than 60 nm for the cell to adhere to the substrate, that cells are able to sense small differences in ligand concentration and move toward greater ligand densities, and that at least 6 integrins must cluster together to form a stable focal adhesion.

4 Conclusion

Experiments such as those presented here have given biologists a clearer picture of the forces found in native biological systems and given them more powerful tools to influence cell behavior. This type of knowledge could be helpful in designing better materials for a variety of biomedical applications and could make it possible to encode signals into surfaces in order to direct cell behavior.

Bibliography

Bruce Alberts, Alexander Johnson, Julian Lewis, Martin Raff, Keith Roberts, and Peter Walter. *Molecular Biology of the Cell.* Garland Science, New York, 5 edition, July 2008.

Marco Arnold, E Ada Cavalcanti-Adam, Roman Glass, Jacques Blümmel, Wolfgang Eck, Martin Kantlehner, Horst Kessler, and Joachim P Spatz. Activation of Integrin Function by Nanopatterned Adhesive Interfaces. *ChemPhysChem*, 5(3):383–388, January 2004.

Marco Arnold, Marco Schwieder, Jacques Bluemmel, E Ada Cavalcanti-Adam, Monica Lopez-Garcia, Horst Kessler, Benjamin Geiger, and Joachim P Spatz. Cell interactions with hierarchically structured nano-patterned adhesive surfaces. *Soft Matter*, 5(1):72–77, January 2009.

Jacques Bluemmel, Nadine Perschmann, Daniel Aydin, Jovana Drinjakovic, Thomas Surrey, Monica Lopez-Garcia, Horst Kessler, and Joachim P Spatz. Protein repellent properties of covalently attached PEG coatings on nanostructured SiO(2)-based interfaces. *Biomaterials*, 28(32):4739–4747, November 2007.

William Brieher, Margaret Coughlin, and Timothy Mitchison. Fascin-mediated propulsion of Listeria monocytogenes independent of frequent nucleation by the Arp2/3 complex. *The Journal of Cell Biology*, 165(2): 233–42, 2004.

Lisa Cameron, Matthew Footer, Alexander van Oudenaarden, and Julie Theriot. Motility of ActA protein-coated microspheres driven by actin polymerization. *Proceedings of the National Academy of Sciences of the United States of America*, 96(9):4908–13, April 1999.

E Ada Cavalcanti-Adam, Alexandre Micoulet, Jacques Bluemmel, Jörg Auernheimer, Horst Kessler, and Joachim P Spatz. Lateral spacing of integrin ligands influences cell spreading and focal adhesion assembly. *European Journal of Cell Biology*, 85(3-4):219–224, January 2006.

E Ada Cavalcanti-Adam, Daniel Aydin, Vera Catherine Hirschfeld-Warneken, and Joachim P Spatz. Cell adhesion and response to synthetic nanopatterned environments by steering receptor clustering and spatial location. *HFSP Journal*, 2(5):276–285, October 2008.

Adam J Engler, Shamik Sen, H Lee Sweeney, and Dennis E Discher. Matrix Elasticity Directs Stem Cell Lineage Specification. *Cell*, 126(4):677–689, August 2006.

Matthew Footer, Jacob Kerssemakers, Julie Theriot, and Marileen Dogterom. Direct measurement of force generation by actin filament polymerization using an optical trap. *Proceedings of the National Academy of Sciences of the United States of America*, 104(7):2181–6, February 2007.

Benjamin Geiger, Alexander Bershadsky, Roumen Pankov, and Kenneth M Yamada. Transmembrane crosstalk between the extracellular matrix-cytoskeleton crosstalk. *Nature Reviews Molecular Cell Biology*, 2(11): 793–805, November 2001.

Filippo G Giancotti and Erkki Ruoslahti. Integrin Signaling. *Science*, 285 (5430):1028–1033, August 1999.

Philippe P Girard, E Ada Cavalcanti-Adam, Ralf Kemkemer, and Joachim P Spatz. Cellular chemomechanics at interfaces: sensing, integration and response. *Soft Matter*, 3:307–326, January 2007.

Vera Hirschfeld-Warneken, Marco Arnold, E Ada Cavalcanti-Adam, M Lopez-Garcia, Horst Kessler, and Joachim P Spatz. Cell adhesion and polarisation on molecularly defined spacing gradient surfaces of cyclic RGDfK peptide patches. *European Journal of Cell Biology*, 87(8-9):743–750, June 2008.

Hayden Huang, Roger D Kamm, and Richart T Lee. Cell mechanics and mechanotransduction: pathways, probes, and physiology. *AJP: Cell Physiology*, 287(1):C1–C11, February 2004.

Jinghuan Huang, Stefan V Grater, Francesca Corbellinl, Sabine Rinck, Eva Bock, Ralf Kemkemer, Horst Kessler, Jiandong Ding, and Joachim P Spatz. Impact of Order and Disorder in RGD Nanopatterns on Cell Adhesion. *Nano Letters*, 9(3):1111–1116, January 2009.

Paul A Janmey. The cytoskeleton and cell signaling: component localization and mechanical coupling. *Physiological Reviews*, 78(3):763–781, July 1998.

Simon Jungbauer, Huajian Gao, Joachim P Spatz, and Ralf Kemkemer. Two Characteristic Regimes in Frequency-Dependent Dynamic Reorientation of Fibroblasts on Cyclically Stretched Substrates. *Biophysical Journal*, 95(7):3470–3478, October 2008.

Torsten Kühne, Reinhard Lipowsky, and Jan Kierfeld. Zipping mechanism for force generation by growing filament bundles. *Europhysics Letters*, 86(6):68002, 2009.

Theobald Lohmüller, Daniel Aydin, Marco Schwieder, Christoph Morhard, Ilia Louban, Claudia Pacholski, and Joachim P Spatz. Nanopatterning by block copolymer micelle nanolithography and bioinspired applications. *Biointerphases*, 6(1):MR1, 2011.

Shingo Miyamoto, Steven K Akiyama, and Kenneth M Yamada. Synergistic roles for receptor occupancy and aggregation in integrin transmembrane function. *Science*, 267(5199):883–885, February 1995.

Alexander Mogilner and Boris Rubinstein. The physics of filopodial protrusion. *Biophysical Journal*, 89(2):782–95, 2005.

Rajalakshmi Nambiar, Russell McConnell, and Matthew Tyska. Myosin motor function: the ins and outs of actin-based membrane protrusions. *Cellular and Molecular Life Sciences*, 67(8):1239–54, 2010.

Charles Peskin, Garret Odell, and George Oster. Cellular motions and thermal fluctuations: the Brownian ratchet. *Biophysical Journal*, 65(1): 316–324, 1993.

Sander Pronk, Phillip Geissler, and Daniel Fletcher. Limits of Filopodium Stability. *Physical Review Letters*, 100(25):4–7, 2008.

Erkki Ruoslahti and Michael D Pierschbacher. New Perspectives in Cell-Adhesion - RGD and Integrins. *Science*, 238(4826):491–497, 1987.

Joachim P Spatz, Stefan Mössmer, Christoph Hartmann, Martin Möller, Thomas Herzog, Michael Krieger, Hans-Gerd Boyen, Paul Ziemann, and Bernd Kabius. Ordered Deposition of Inorganic Clusters from Micellar Block Copolymer Films. *Langmuir*, 16(2):407–415, January 2000.

Martin Streichfuss, Friedrich Erbs, Kai Uhrig, and Rainer Kurre. Measuring Forces between Two Single Actin Filaments during Bundle Formation. *Nano Letters*, 11(9):3676–80, September 2011.

Tatyana Svitkina, Elena Bulanova, and Oleg Chaga. Mechanism of filopodia initiation by reorganization of a dendritic network. *The Journal of Cell Biology*, 160:409–421, 2003.

Manuel Théry, Anne Pépin, Emilie Dressaire, Yong Chen, and Michel Bornens. Cell distribution of stress fibres in response to the geometry of the adhesive environment. *Cell Motility and the Cytoskeleton*, 63(6): 341–355, June 2006.

Kai Uhrig, Rainer Kurre, Christian Schmitz, Jennifer E Curtis, Tamás Haraszti, Anabel E-M Clemen, and Joachim P Spatz. Optical force sensor array in a microfluidic device based on holographic optical tweezers. *Lab on a chip*, 9(5):661–8, March 2009.

Eli Zamir and Benjamin Geiger. Molecular complexity and dynamics of cell-matrix adhesions. *Journal of Cell Science*, 114(20):3583–3590, 2001.

Microfluidic Platforms for Human Disease Cell Mechanics Studies

Ebrahimi Warkiani Majid [1] and Chwee Teck Lim [1,2,3]

[1] Singapore-MIT Alliance for Research Technology Center, Singapore
[2] Department of Bioengineering, National University of Singapore, Singapore
[3] Mechanobiology Institute, National University of Singapore, Singapore

Introduction

Microfluidics is an interdisciplinary field at the interface of chemistry, engineering, and biology; and has experienced rapid growth over the past decades due to advantages associated with miniaturization, integration and faster sample processing and analysis time (Gervais et al., 2011; Hou et al., 2011; Bhagat et al., 2010). Recently, several microfluidic platforms have been developed for the study of human disease cell biomechanics at the cellular and molecular levels so as to gain better insights into various human diseases such as cancer (Bhagat et al., 2010), pneumonia (Kim et al., 2009b), sepsis (Mach and Di Carlo, 2010) and malaria (Hou et al., 2010). In this section, we will elaborate on recent advances in cellular biomechanics using microfluidic approaches. In particular, we will look at various techniques in probing cellular mechanical properties with some novel applications in cancer and malaria such as theidentification and enrichments of these diseased cells from their normal counter parts. We will also provide insights into the challenges associated with current microfluidic approaches and provide future perspectives for the next-generation platforms.

Investigating human disease cell mechanics

Mechanical stimulation of biological species at various spatial and temporal scales has a direct effect on their function and characteristics such as cell fate, genetic expression, migration, differentiation and adhesion among others (Kim et al., 2009a). On the other hand, such quantitative monitoring and measurements can also reveal valuable information about cell stiffness, heterogeneity and even malignancy (e.g. in the case of tumour cells) as well as genetic alterations at subcellular levels (Hou et al., 2011; Kim et al.,

2009a; Fletcher and Mullins, 2010). Understanding the interplay between cell mechanics and functions (or pathophysiological functions) can also help in the development of robust frameworks towards diagnosis and prognosis of many diseases. There has been growing evidence of the correlation between changes in cell mechanical properties and disease progression (Wirtz et al., 2011). For instance, tumor cells with higher malignancy are known to be more deformable than benign epithelial cells, and this may give rise to their ability to metastasize and eventually extravasate into distant organs during disease progression. In the case of infectious diseases such as malaria, the red blood cells (RBCs) become stiff and cytoadhesive when parasites invade and mature within the RBCs (Kim et al., 2009a; Zare and Kim, 2010).

Microfluidics is emerging as an enabling platform technology for cell mechanobiology studies, primarily due to its small length scale, reduced sample and reagent volumes and well-developed microfabrication techniques (Hou et al., 2011). Employment of biocompatible and transparent materials such as glass and polymers (i.e., COC, PDMS and PMMA) with advanced microfabrication techniques has enabled researchers to develop various miniaturized devices for accurate and quantitative analysis of cellular responses to mechanical (or chemical) stimulus with precise control of the cellular environment (Franke and Wixforth, 2008).

Microfabrication technique

Microfabrication is a process to construct miniature structures with micrometer size or smaller. Over the past decades, microfabrication has been the workhorse behind the phenomenal growth in many industries such as microelectronics, pharmaceuticals and the automotive (Vogelaar, 2005; Madou, 1997). Microfabrication has also supported many new technologies such as micro/nanofluidics, Bio-MEMS, photonics, and biosensors where miniaturization and large-scale integration are required for a reduction in time, costs, sample size, or power consumption as well as portability and reliability (Vogelaar, 2005; Voldman et al., 1999). In the areas where microfabrication is applied, methods to realize structures on such small length scales can be divided in direct writing methods (and/or photolithography) and replication methods (Warkiani, 2012).

Direct writing methods in the semiconductor industry refer to those techniques that can be combined directly to CAD/CAM software to generate micrometer patterns for the manufacturing of microelectronic devices and other micro-engineered components such as MEMS and microfluidic devices (Zhang et al.). For example, writing with a focused laser beam is normally used to fabricate photomasks that contain the structure necessary for conventional photo-mask lithography or can be employed directly for micro/nano patterning (e.g., electron beam (e-beam) and focused ion-beam (FIB) lithography). Photolithography is a process used widely in micro-fabrication to transfer the desired pattern from a photo-mask to a film of photoresist on the substrate (Madou, 1997; Geissler and Xia, 2004). The patterns generated by lithography are then transferred into the underlying material by etching (i.e., wet or dry etching), or transferred into other layers by deposition. The substrate materials that are normally used in the semiconductor industry are mainly silicon, glass, quartz, metals, and some polymers such as PMMA (Poly(methyl methacrylate)) and PC (Polycarbonate) (Vogelaar, 2005; Qin et al., 1998). The limited transparency of these materials, and the high costs are the major drawbacks for the use of photolithography in mechanobiology and microfluidics. In addition, the inability to pattern non-planar surfaces as well as limitations in geometrical design and surface chemistry caused researchers to look for alternative approaches, where micromolding methods emerged (Becker and Locascio, 2002). Micromolding relies on replication of a master mold having a micro/nano structured relief profile on its surface, and can be divided to following categories: (i) Soft-lithography and (ii) Micromolding against a rigid mold.

Soft lithography refers to a collection of techniques for fabricating or replicating micro/nano structures using elastomeric stamps based on printing, embossing or molding (Xia and Whitesides, 1998; Weibel et al., 2007). The master mold mostly is prepared from a soft material (i.e., polydimethylsiloxane (PDMS)), by curing on a micro-fabricated mold (see Fig. 1A).

Soft-lithography has some advantages over others such as lower cost in mass production, being ideal for biotechnology applications and the ability to transfer pattern on non-planar surfaces (Weibel et al., 2007). Soft-lithography techniques, which mainly developed within the group of G.M. Whitesides at Harvard University (Xia et al., 1996) can be classified into six main groups: microcontact printing, replica molding, micromolding in capillaries, capillary force lithography, microtransfer molding and solvent assisted micromolding (Xia et al., 2004). Among them, replica and microtransfer molding are the most important categories of soft-lithography that have found the most practical applications in cell biomechanics, microbiology and synthetic biology (Whitesides, 2006; Li et al., 2003). In these techniques, the shape, size and pattern of features on a master mold (i.e., Silicon/glass mold which is normally made by conventional lithography and microfabrication techniques) is transferred to the PDMS stamps by soft-lithography. This intermediate stamps can be used directly for fabrication of microchannels (by bonding to a glass/polymeric substrate) and

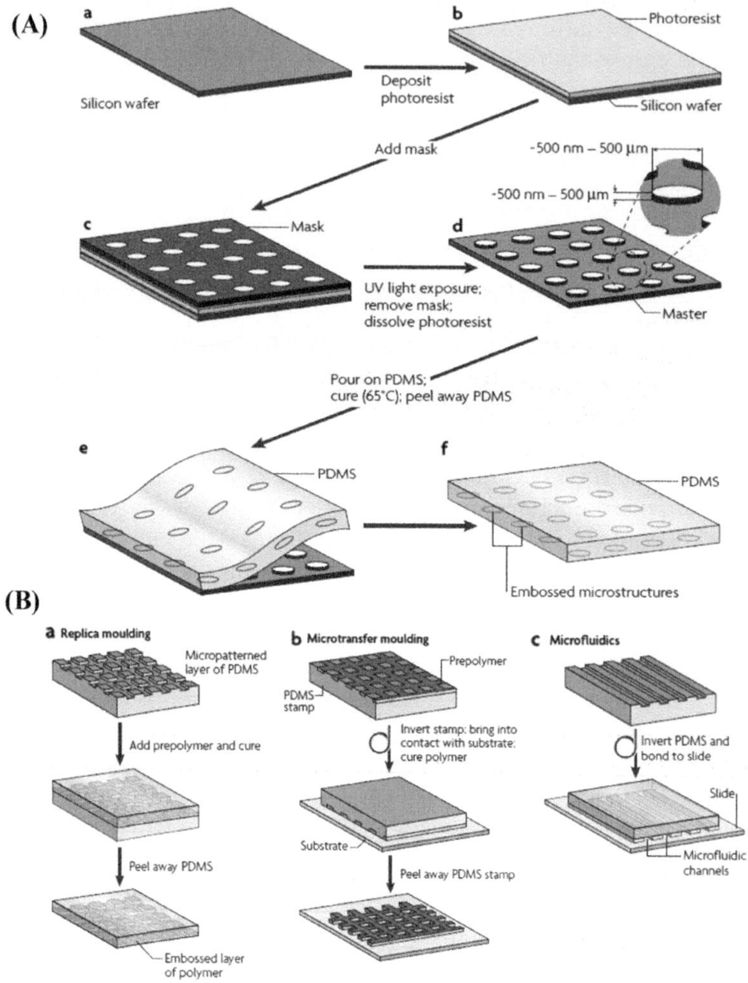

Figure 1. (A) Schematic of the fabrication of PDMS stamp for soft-lithography: (a)-(c) photoresist is spin-coated on a silicon substrate and a mask with desired pattern is placed in contact with the layer of photoresist, (d) photolithography is carried out using an ultraviolet (UV) light through the mask and pattern is achieved by dissolving the non-crosslinked resists inside a developer, (e) uncured PDMS is poured on the master and cured thermally and peeled away, and (f) the resulting layer of PDMS has microstructures embossed in its surface. (B) The schematic representation of the major soft-lithography techniques: (a) replica molding, (b) micro-transfer molding, and (c) microfluidics. Reprinted with permission from Weibel et al. (2007), copyright 2007.

cell patterning, or can be employed as a master mold again for replication of another replica to micropattern biocompatible polymers, including epoxies, polyurethanes, polyethylene glycol (PEG), agar and agarose (see Fig. 1B) (Weibel et al., 2007).

Cancer cell mechanics

Cancer is the leading cause of death worldwide, accounting for 7.6 million deaths in 2008 (Allan and Keeney, 2009). One defining feature of cancer is the rapid creation of abnormal cells that grow beyond their usual boundaries, and which can then shed into blood stream and spread to other organs (i.e. metastasis) (Alix-Panabires et al., 2012). The increase in deformability of circulating cells is emerging as an important biomarker for the diagnostic and prognostic of patients with cancer (Vaziri and Gopinath, 2007). In particular, the dynamic reorganization of the cytoskeleton has become a specific point of interest regarding changes in cell characteristics and functions (Fletcher and Mullins, 2010).

Capillary-based microfluidic techniques have recently attracted much attention for investigating the biomechanics of cancer cells (Kim et al., 2009a). The constricted microchannel inside the microfluidic devices with precisely defined geometries allows single cells to squeeze under pressure and mimic the *in vitro* capillary-like microenvironment to study the biorheological behavior of cells as they passage through narrow constrictions of the blood capillaries (Suresh, 2007). For example, Hou et al., (Hou et al., 2009) employed a simple microfluidic platform with a straight square microchannel (1010m) to measure the deformability of cancerous breast cells (MCF-7) in comparison to the benign breast epithelial cells (MCF-10A). By measuring some parameters such as entry time, elongation index and transit velocity, they could clearly distinguish the difference in stiffness of cancerous and benign breast cells. In a similar study, Adamo and colleagues (Adamo et al., 2012) demonstrated a high throughput microfluidics system to probe deformability of single cells with up to 800 cells/min. They showed that cell travel time through a funnel-shaped constriction has a non-Gaussian distribution. Besides, they found that cells with different diameters have different travel times and that, for a given diameter, stiffer cells have longer transit times than less stiff ones. In another investigation, a microfluidic micropipette aspiration system was developed by Gou and colleages (Guo et al., 2012a) for measuring the deformability of single cells at different pressures. They could precisely measure the deformability of several types of nucleated cells such as human neutrophils, lymphocytes and RT4 bladder cancer cells by leveraging the advantages of their microfluidic platform over conventional micropipette aspiration systems. For better understanding of cancer metastasis, Perez et al., (Gallego-Perez et al., 2012) recently developed a guided cell migration microfluidic platform to mimic the fiber-like structure of stroma (or parenchyma) using a micro-patterned polystyrene substrate. They studied the migratory behavior of primary (glioma) and metastatic (lung and colon) tumors at single-cell level and found that all the tumor cells exhibited axially persistent cell migration where lung tumor cells showed the highest migratory velocities compared to glioma and colon tumor cells. Gabriele et al. (Gallego-Perez et al., 2012; Gabriele et al., 2009) studied the role of cytoskeletal components such as actin and myosin II in capillary leukocyte trafficking through an array of interconnected capillary segments, which is the starting event of trauma disorders in the lung microvasculature. They showed that the redistribution of actin filaments has distinct roles in the extent and rate of cell deformation and cell membrane unfolding at the different stages (i.e., the entry stage, transit stage, and shape recovery stage). In addition, they demonstrated that cell shape and its orientation at the channel entrance can directly affect the overall transit time measurement. Preira et al., (Preira et al., 2012) recently developed a microfluidic system made of successive comb-like filters in series with different pore size to separate circulating leukocytes based on deformability. Their results show that stiffening of whole leukocyte population affects patients with acute respiratory distress syndrome (ARDS) while previous data from Rosenbluth et al., (Rosenbluth et al., 2008)indicate that only a mere fraction of leukocytes, around 5%, have an altered stiffness. Using the same concept, Zhang et al., (Zhang et al., 2012) reported a microfluidics method to enrich physically flexible cells, which are cells with high metastatic propensity (i.e., cancer stem cells) from other counterparts (i.e., stiff cells) through the micro-barriers (see Fig. 2). They found a good correlation between the phenotype of deformable cells that passed through the micro-barriers and over expression of multiple genes, which are involved in cancer cell motility and metastasis.

Malaria infected red blood cell mechanics

Malaria is one of the deadliest human infectious diseases with over a billion individuals at risk of infection. The prime causal parasite, *Plasmodium (P.) falciparum*, is responsible for about 2 million deaths a year (Hou et al., 2011; Handayani et al., 2009). *P. falciparum*-infected red blood cells (iRBCs) undergo various developmental stages in which during the last stage, they lose their biconcave shape, become more spherical, and exhibit elevated stiffness [6]. This loss of deformability is often reported to be the

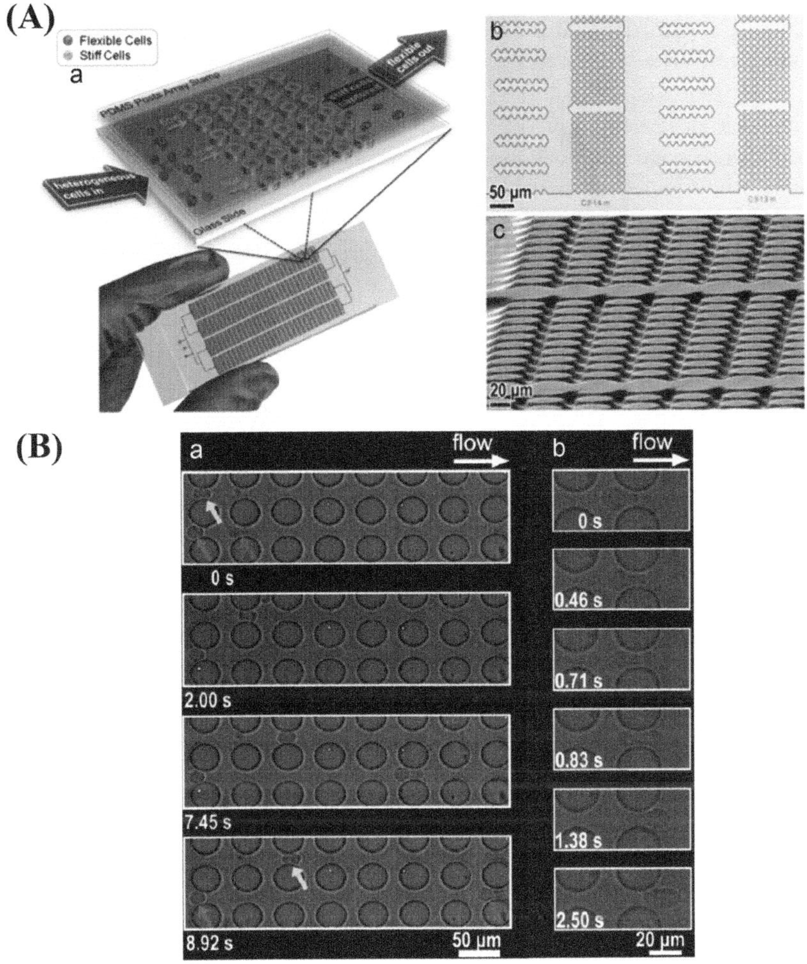

Figure 2. (A-a) Schematic and photograph of a mechanical separation (MS) chip. (A-b) Optical microscopic images of the MS-chip. (A-c) SEM image of the micro-pillars used for cell separation. (B-a) Optical images of SUM149 cells crossing the MS-chip gaps. Red, green, and yellow arrows indicate different levels of cell deformability. The most flexible cell type (red) passes the entire micro-pillar array; the less flexible one (yellow) crosses halfway; and the least flexible one (green) stops at the initial intersection. (B-b) High-magnification images show cell deformation in different time frames. Reprinted with permission from Zhang et al. (Zhang et al., 2012), copyright 2012.

main factor in capillary blockage (Shelby et al., 2003). Similar to previous section, several microfluidic devices have been developed to study the behavior of stiffer iRBCs under capillary microenvironments. In 2003, Shelby et al. (Shelby et al., 2003) employed elastomeric microfluidic chips to study biorheological behaviour of iRBCs *ex vivo* by passing iRBCs of different stages into small microchannels to characterize complex behaviors of single cells, under flow, in multicellular capillary blockages. Then showed that stiffer late stage iRBC scan easily block the small microchannels with 24 m width. They also observed the pitting phenomena where spleen beds remove parasites without destroying the RBCs inside microchannel with 2 m width. More recently, Bow and colleagues (Bow et al., 2011) utilized the same concept to fabricate a deformability cytometer that measures dynamic mechanical responses of 103104 individual RBCs in a heterogeneous cell population (Fig. 3A). They demonstrated the ability of their system for high-throughput characterization and separation of iRBCs from uninfected ones. Quantitative biomechanical characteristics of individual RBCs from the experimental results was also inferred using a dissipative particle dynamics (DPD) model. In another attempt for characterization of parasitized RBCs, Guo et al., (Guo et al., 2012b) developed a microfluidic platform with funnel-shaped constrictions to measure RBCs deformability using precisely controlled pressure. They demonstrated that parasitized cells from ring through schizont stages were 1.5 to 200 times stiffer than uninfected cells which were in good agreement with previous studies. Their results clearly demonstrated the versatility of their approach for various applications such as functional studies and high-throughput drug screening. Zheng et al. (Zheng et al., 2012) also recently developed a high-throughput microfluidic system for biophysical characterization of RBCs at a speed of 100150 cells s1, which can be used eventually for measurement of multiple biophysical disorders in RBC of patients with sepsis, malaria, or sickle cell anemia. Another study (Handayani et al., 2009) using an array of microfluidic channels also confirmed the biophysical differences between uninfected and *P. falciparum* infected as well as *P.vivax* infected RBCs (Fig. 3B).

Besides investigating cell deformability, microfluidic devices can also be used for separation of iRBCs from other blood cells. Many of these devices are inspired by the *in vivo* physiological cell interactions in spleen and blood vessels such as passage from narrow constrictions or margination. For instance, both ring-stage iRBCs were separated (based on mechanical deformability) from normal RBCs in a continuous-flow fashion using a successive microfilter with 2.2 m gaps (MAJ, 2009). More recently, Hou et al. (Hou et al., 2010) employed a long straight channel microfluidic device to separate iRBCs from whole blood using the effect of margination. In their system, cell margination was directed along the channel width with the iRBCs aligning near each sidewall and then subsequently removed using a 3-outlet system. They demonstrated recovery of around 75% for early stage iRBCs ¿90% for late stage iRBCs at the side outlets.

Conclusions

Living cells are complex and dynamic systems which are inherently sensitive to a myriad of biomechanical cues. Understanding the coupling between the biological complexity and biomechanical properties and responses of cells requires advanced tools for manipulation and testing. Microfluidic platforms enable quantitative analysis of single-cell stimulation and mechanical response measurements and open a new paradigm for the characterization and enrichment of both healthy and diseased cell populations with great accuracy. In particular, microfluidic systems can be employed in cancer for the identification and separation of tumorigenic cells or can be utilized to couple chemicals gradients to shear stress assays for drug discovery. They can also be utilized as a screening/diagnostic tool to explore connections between certain diseases such as malaria or sickle cell anemia and cell stiffness.

In summary, the development of novel platforms is crucial to gain new insights into the cell mechanical property changes and human disease states. While some microfluidic devices can successfully operate with high capacity and various functionality, there are challenges that remain to be solved such as increasing throughput, on-line data analysis, as well as integration of mechanical, optical and electrical stimulation parts within a chip.

Bibliography

A. Adamo, A. Sharei, L. Adamo, B.K. Lee, S. Mao, and K.F. Jensen. Microfluidics-based assessment of cell deformability. *Analytical Chemistry*, 84(15):6438 6443, 2012.

C. Alix-Panabires, H. Schwarzenbach, and K. Pantel. Circulating tumor cells and circulating tumor dna. *Annual Review of Medicine*, 63:199–215, 2012.

A.L. Allan and M. Keeney. Circulating tumor cell analysis: technical and statistical considerations for application to the clinic. *Journal of oncology*, 2010, 2009.

H. Becker and L.E. Locascio. Polymer microfluidic devices. *Talanta*, 56(2): 267–287, 2002.

A.A.S. Bhagat, H. Bow, H.W. Hou, S.J. Tan, J. Han, and C.T. Lim. Mi-

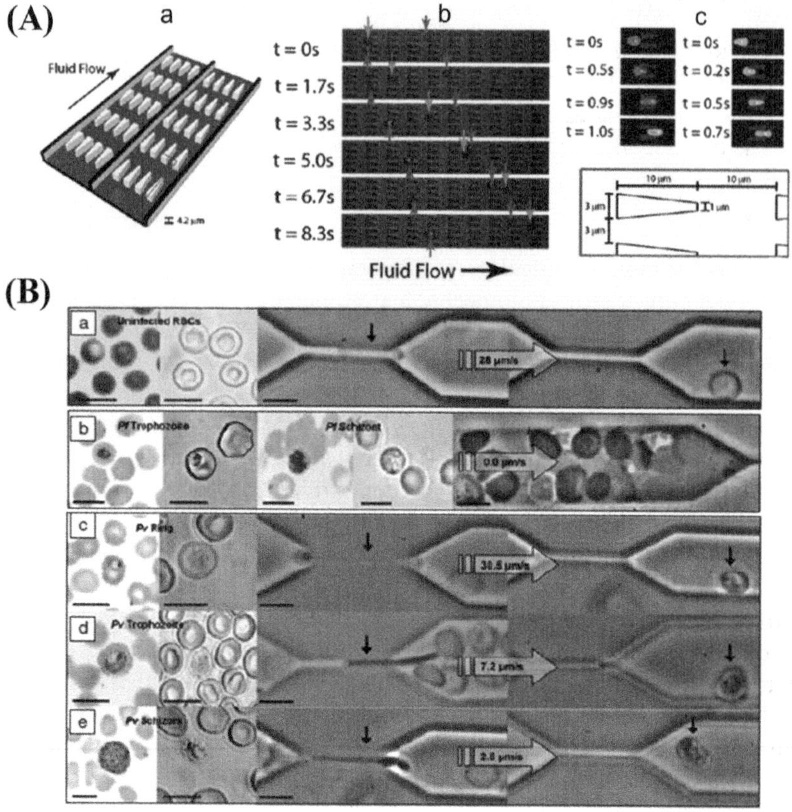

Figure 3. (A-a) Schematic illustration of a microfluidic chip with successive filters. (A-b) Experimental images of ring stage P. falciparum-infected (red arrows) and uninfected (blue arrows) RBCs in the channels. (A-c) Dissipative particle dynamics (DPD) simulation images of P. falciparum-infected RBCs travelling in channels of converging (left) and diverging (right) pore geometry. Reprinted with permission from Bow et al.(Bow et al., 2011), copyright 2011. (B) Passage of (a) uninfected, (b) trophozoite, and schizont stages of P. falciparuminfected RBCs and (ce) ring, trophozoite, and schizont stages of P. vivaxinfected RBCs through 2 m constricted channels. Scale bar = 10 m in all cases. Reprinted with permission from Handayani et al. (Handayani et al., 2009), copyright 2009.

crofluidics for cell separation. *Medical and Biological Engineering and Computing*, 48(10):999–1014, 2010.

H. Bow, I.V. Pivkin, M. Diez-Silva, S.J. Goldfless, M. Dao, J.C. Niles, S. Suresh, and J. Han. A microfabricated deformability-based flow cytometer with application to malaria. *Lab on a Chip*, 11(6):1065–1073, 2011.

D.A. Fletcher and R.D. Mullins. Cell mechanics and the cytoskeleton. *Nature*, 463(7280):485–492, 2010.

T.A. Franke and A. Wixforth. Microfluidics for miniaturized laboratories on a chip. *ChemPhysChem*, 9(15):2140–2156, 2008.

S. Gabriele, A.M. Benoliel, P. Bongrand, and O. Thodoly. Microfluidic investigation reveals distinct roles for actin cytoskeleton and myosin ii activity in capillary leukocyte trafficking. *Biophysical journal*, 96(10):4308–4318, 2009.

D. Gallego-Perez, N. Higuita-Castro, L. Denning, J. DeJesus, K. Dahl, A. Sarkar, and D.J. Hansford. Microfabricated mimics of in vivo structural cues for the study of guided tumor cell migration. *Lab on a Chip*, 2012.

M. Geissler and Y. Xia. Patterning: Principles and some new developments. *Advanced Materials*, 16(15):1249–1269, 2004.

L. Gervais, N. De Rooij, and E. Delamarche. Microfluidic chips for pointofcare immunodiagnostics. *Advanced Materials*, 23(24):H151–H176, 2011.

Q. Guo, S. Park, and H. Ma. Microfluidic micropipette aspiration for measuring the deformability of single cells. *Lab Chip*, 2012a.

Q. Guo, S.J. Reiling, P. Rohrbach, and H. Ma. Microfluidic biomechanical assay for red blood cells parasitized by plasmodium falciparum. *Lab on a Chip*, 12(6):1143–1150, 2012b.

S. Goldfless P. Abgrall K. Tan J. Niles C. T. Lim J. Han H. Bow, H. W. Hou. Continuous-flow deformability-based sorting of malaria-infected red blood cells. *mTAS 2009*, Jeju, Korea, pp. 1219, 2009.

S. Handayani, D.T. Chiu, E. Tjitra, J.S. Kuo, D. Lampah, E. Kenangalem, L. Renia, G. Snounou, R.N. Price, and N.M. Anstey. High deformability of plasmodium vivax-infected red blood cells under microfluidic conditions. *Journal of Infectious Diseases*, 199(3):445–450, 2009.

HW Hou, QS Li, G.Y.H. Lee, AP Kumar, CN Ong, and CT Lim. Deformability study of breast cancer cells using microfluidics. *Biomedical microdevices*, 11(3):557–564, 2009.

H.W. Hou, A.A.S. Bhagat, A.G.L. Chong, P. Mao, K.S.W. Tan, J. Han, and C.T. Lim. Deformability based cell marginationa simple microfluidic design for malaria-infected erythrocyte separation. *Lab Chip*, 10(19):2605–2613, 2010.

H.W. Hou, W.C. Lee, M.C. Leong, S. Sonam, S.R.K. Vedula, and C.T. Lim. Microfluidics for applications in cell mechanics and mechanobiology. *Cellular and Molecular Bioengineering*, pages 1–12, 2011.

D.H. Kim, P.K. Wong, J. Park, A. Levchenko, and Y. Sun. Microengineered platforms for cell mechanobiology. *Annual review of biomedical engineering*, 11:203–233, 2009a.

K. Kim, H.S. Jung, J.Y. Song, M.R. Lee, K.S. Kim, and K.Y. Suh. Rapid detection of mycoplasma pneumonia in a microfluidic device using immunoagglutination assay and static light scattering. *Electrophoresis*, 30 (18):3206–3211, 2009b.

N. Li, A. Tourovskaia, and A. Folch. Biology on a chip: microfabrication for studying the behavior of cultured cells. *Critical reviews in biomedical engineering*, 31(5-6):423–488, 2003.

A.J. Mach and D. Di Carlo. Continuous scalable blood filtration device using inertial microfluidics. *Biotechnology and bioengineering*, 107(2): 302–311, 2010.

M. Madou. *Fundamentals of microfabrication*, volume 1. CRC Press LLC. Captulo, 1997.

P. Preira, V. Grandne, M. Camara, S. GABRIELE, J.M. Forel, and O. Theodoly. Passive circulating cell sorting by deformability using a microfluidic gradual filter. *Lab on a Chip*, 2012.

D. Qin, Y. Xia, A.J. Black, and G.M. Whitesides. Photolithography with transparent reflective photomasks. *Journal of Vacuum Science Technology B: Microelectronics and Nanometer Structures*, 16(1):98–103, 1998.

M.J. Rosenbluth, W.A. Lam, and D.A. Fletcher. Analyzing cell mechanics in hematologic diseases with microfluidic biophysical flow cytometry. *Lab Chip*, 8(7):1062–1070, 2008.

J.P. Shelby, J. White, K. Ganesan, P.K. Rathod, and D.T. Chiu. A microfluidic model for single-cell capillary obstruction by plasmodium falciparum-infected erythrocytes. *Proceedings of the National Academy of Sciences*, 100(25):14618–14622, 2003.

S. Suresh. Biomechanics and biophysics of cancer cells. *Acta Materialia*, 55 (12):3989–4014, 2007.

A. Vaziri and A. Gopinath. Cell and biomolecular mechanics in silico. *Nature materials*, 7(1):15–23, 2007.

L. Vogelaar. Phase separation micromolding. phd dissertation, university of twente, enschede, the netherlands, 2005.

J. Voldman, M.L. Gray, and M.A. Schmidt. Microfabrication in biology and medicine. *Annual review of biomedical engineering*, 1(1):401–425, 1999.

M.E. Warkiani. Fabrication and characterization of micro/nano filter for isolation of waterborne pathogens. phd dissertation, nanyang technological university, singapore, 2012.

D.B. Weibel, W.R. DiLuzio, and G.M. Whitesides. Microfabrication meets microbiology. *Nature Reviews Microbiology*, 5(3):209–218, 2007.

G.M. Whitesides. The origins and the future of microfluidics. *Nature*, 442 (7101):368–373, 2006.

D. Wirtz, K. Konstantopoulos, and P.C. Searson. The physics of cancer: the role of physical interactions and mechanical forces in metastasis. *Nature Reviews Cancer*, 11(7):512–522, 2011.

Y. Xia and G.M. Whitesides. Soft lithography. *Annual review of materials science*, 28(1):153–184, 1998.

Y. Xia, D. Qin, and G.M. Whitesides. Microcontact printing with a cylindrical rolling stamp: A practical step toward automatic manufacturing of patterns with submicrometersized features. *Advanced Materials*, 8(12): 1015–1017, 1996.

Y. Xia, J.J. McClelland, R. Gupta, D. Qin, X.M. Zhao, L.L. Sohn, R.J. Celotta, and G.M. Whitesides. Replica molding using polymeric materials: A practical step toward nanomanufacturing. *Advanced Materials*, 9(2):147–149, 2004.

R.N. Zare and S. Kim. Microfluidic platforms for single-cell analysis. *Annual review of biomedical engineering*, 12:187–201, 2010.

W. Zhang, K. Kai, D.S. Choi, T. Iwamoto, Y.H. Nguyen, H. Wong, M.D. Landis, N.T. Ueno, J. Chang, and L. Qin. Microfluidics separation reveals the stem-celllike deformability of tumor-initiating cells. *Proceedings of the National Academy of Sciences*, 2012.

Y. Zhang, C. Liu, and D. Whalley. Direct-write techniques for maskless production of microelectronics: A review of current state-of-the-art technologies. In *Electronic Packaging Technology High Density Packaging, 2009. ICEPT-HDP'09. International Conference on*, pages 497–503. IEEE.

Y. Zheng, E. Shojaei-Baghini, A. Azad, C. Wang, and Y. Sun. High-throughput biophysical measurement of human red blood cells. *Lab on a Chip*, 2012.

Continuum analyses of structures containing cracks

Roberto Ballarini
Department of Civil Engineering, University of Minnesota, Minneapolis, USA

1 Introduction

The previous chapters of this book have demonstrated how multi-scale models and experiments at very small scales have improved and offer promise for furthering our understanding of the mechanical behavior and physical properties of biological materials and structures. This chapter illustrates how valuable insights can be gained using simplified continuum mechanics models that gloss over the complexities that can be handled by the multi-scale models, such as numerous distinct nano and micro scale features that comprise the structures, material inhomogeneity, biological and chemical processes, the precise geometrical description, and the presence of flaws. This as long as the simplified treatments capture the essential features controlling the mechanical response. The reward for sacrificing the details is that the simplified models are amenable to analytical treatment. The discussion is limited to linear elastic structures containing cracks. The fracture mechanics analyses rely on the concepts associated with the energy release rate (ERR), the universal form of the stress and displacement components in the vicinity of the crack front, the generalized Griffith crack growth criterion for small-scale-yielding (SSY) conditions, and stability of crack propagation under displacement and load controls. Using numerous examples the point will be driven home that if one can accurately estimate the changes in stored elastic energy produced by variations in the crack length, then conditions for crack extension and the stability of crack propagation can be established. The ERR represents the variation in strain energy, or alternatively in potential energy resulting from crack extension. It is often referred to as the crack driving force because as discussed subsequently a crack will extend only if the ERR is sufficiently large to overcome the material's resistance to the creation of the additional crack surfaces. In addition, for the perfectly brittle materials considered here the stability of crack propagation is determined by the change in the ERR with respect to a unit increase in crack surface area. All derivations and examples are for planar and traction-free cracks that grow along their defining plane. Two definitions of ERR are provided first, both motivated from the load-displacement curves of a structure containing cracks having slightly different lengths. Selected examples are then presented and solved that elucidate how accurate approximations of the ERR can be achieved for certain types of crack configurations using simplified stress analysis procedures. This is followed by relating the ERR to the singular stress field in the immediate vicinity of the crack front (in terms of the to-be defined stress intensity factors) and to the global stiffness of the cracked component.

2 Energy approach to fracture

2.1 Definitions of ERR

Consider the planar traction-free crack in the three-dimensional body shown in Figure 1. Denote the footprint of the crack and the perimeter that defines its front by S_{cr} and Σ_{cr}, and the bounding surfaces of the solid with prescribed displacements, u_i^o, or prescribed traction, t_i^o, by S_u and S_t. Without loss of generality S_{cr}, which is the projection of the top and bottom surfaces of the crack, $S^+$and S^-, is chosen to lie in the 1-3-plane. Consider the response of the cracked solid under the action of generalized forces Q_i and generalized conjugate displacements q_i. The elastic energy stored within it depend not only on the displacements, but also on the size of the crack. It is therefore written as $U(q_i, S_{cr})$. If the crack does not extend then the increment of work performed by the forces is stored as additional elastic energy according to the principle of virtual work

$$\delta U \equiv \int_V \sigma_{ij}\delta\varepsilon_{ij}dV = Q_i\delta q_i \equiv \int_S t_i\delta u_i dS + \int_V f_i\delta u_i dV \tag{1}$$

The two crack surfaces represent an additional boundary of the solid, so that in 1, $S \equiv S_t \cup S_u \cup S^+ \cup S^-$. If the crack-front extends by an amount δl that can vary along Σ_{cr}, then the original crack surface area increases by the amount $\delta S_{cr} = 2\int_{\Sigma_{cr}} \delta l d\Sigma_{cr}$. The additional surface area renders the body less stiff, so that it will store less elastic energy for a given displacement. The ERR, G, is defined as the reduction in strain energy resulting from the creation of the additional crack surface area

$$\delta U = Q_i\delta q_i - \int_{\Sigma_{cr}} 2G\delta l d\Sigma_{cr} \tag{2}$$

Equation 2 shows that the ERR is simply the generalized force corresponding to the variation in crack area, in the same way that the Q_i are the

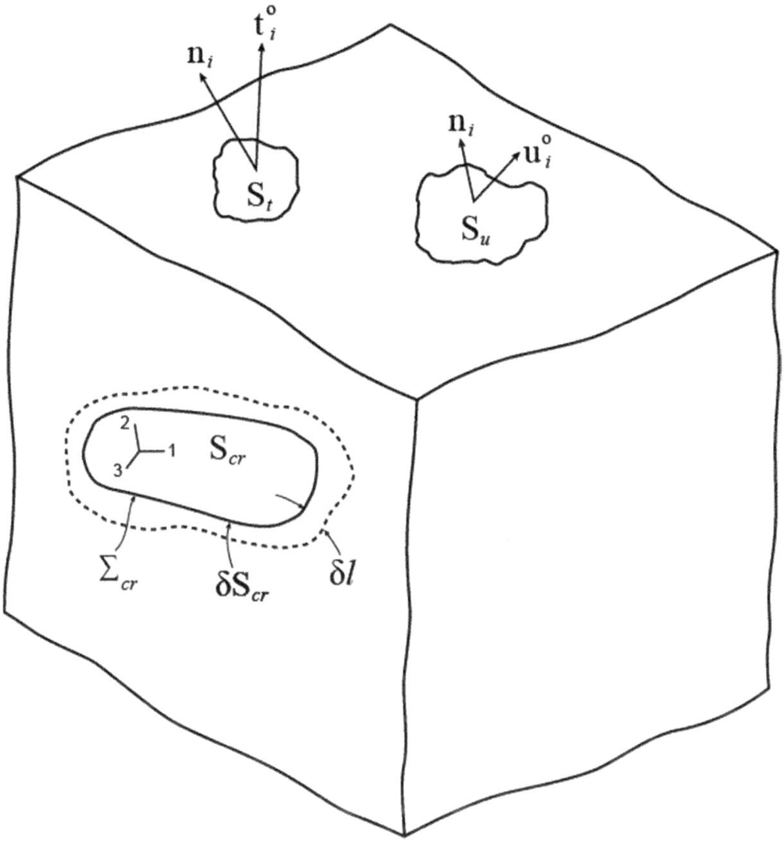

Figure 1. Extension of a planar crack in a three-dimensional solid.

generalized forces associated with the variation in displacements q_i .

The Legendre Transformation of Equation 2 gives the potential energy, $\Pi(Q_i, S_{cr}) \equiv U - Q_i q_i$, a function of the generalized forces and crack area, and whose variation is

$$\delta\Pi = \delta(U - Q_i q_i) = \delta U - q_i \delta Q_i - Q_i \delta q_i \tag{3}$$

Substitution of Equation 2 into 3 provides the alternative definition of the ERR

$$\delta\Pi = -q_i \delta Q_i - \int_{\Sigma_{cr}} 2G \delta l d\Sigma_{cr} \tag{4}$$

The ERR can therefore be defined as the area rate of change of either the strain energy or the potential energy. As will be explained subsequently, which definition is more convenient depends on whether one is analyzing crack extension under constant load or constant displacement conditions. The motivation of the two definitions of ERR comes from considerations of the changes in strain energy and potential energy of a plate of thickness b (Figure 2) containing an initial crack of length l that extends by δl, so that $\delta S_{cr} = 2b\delta l$. These changes are shown in the load-displacement curves that would be produced by testing two replicas of the plate sketched in Figure 2, one containing an initial crack of length l, the other containing an initial crack of length $l + \delta l$. As the load on the plate containing the shorter crack is increased, the load-displacement curve follows path
$0-(Q_1, q_1)$. If the crack extends at Q_1, then the stiffness of the structure decreases, and the response shifts to a point (Q_2, q_2) that lies somewhere on the load-displacement curve corresponding to the specimen containing the longer crack. The coordinates (Q_2, q_2) depend on the changes in force and displacement that may occur during the extension.

Fixed grips corresponds to the case for which the displacement q remains constant during crack extension, so that the load displacement curve of the shorter initial crack follows path $(Q_1, q_1) - (Q_3, q_1)$. The corresponding reduction in strain energy, represented by the shaded area defined by points $0 - (Q_1, q_1) - (Q_3, q_1)$, is $[U(l + \delta l) - U(l)]_{fixed\,q} = \delta U_{fixed\,q}$. For this case $\delta q_i = 0$, and Equation 2 reduces to

$$\delta U_{fixed\ q_i} = -\int_{\Sigma_{cr}} 2G\ \delta\ell\ d\Sigma_{cr} \tag{5}$$

or

$$G_{fixed\ q_i} = -\frac{\partial U}{\partial S_{cr}} \tag{6}$$

which for the plate shown in Figure 2 reduces to $G_{fixedq} = -\frac{1}{b}\frac{\partial U}{\partial l}$. For the fixed grips condition the load does no work during crack extension, so that as discussed subsequently only the release of strain energy can contribute to overcoming the material's resistance to crack propagation.

Under *dead load* conditions, the load remains constant during crack growth, and the load-displacement curve follows path $(Q_1, q_1) - (Q_1, q_3)$. For this case the decrease in potential energy, which is equal to negative the increase in complementary energy, $[\Pi(l + \delta l) - \Pi(l)]_{fixed\,q} = \delta\Pi_{fixed\,q}$ is represented by the sum of the areas defined by points $0 - (Q_1, q_1) - (Q_3, q_1)$ and the points $(Q_1, q_1) - (Q_1, q_3) - (Q_3, q_1)$. Note that the latter contribution becomes negligible as $\delta l \to 0$, and therefore for infinitesimal crack

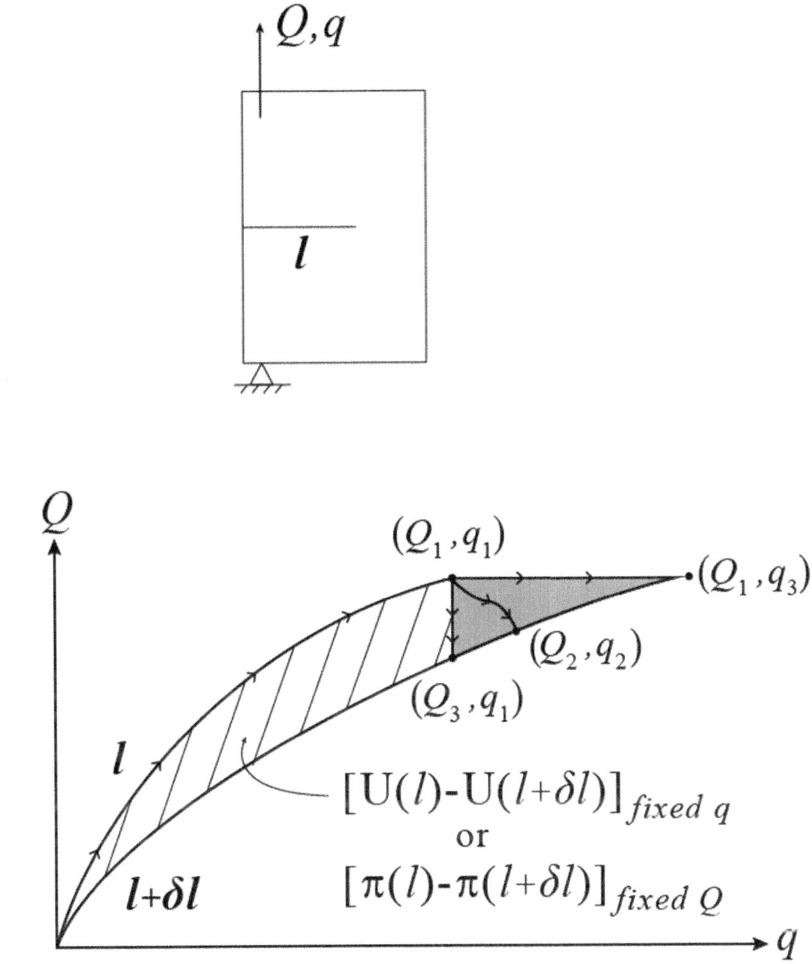

Figure 2. Changes in strain energy and complementary energy with respect to crack extension.

extension the same ERR is obtained for fixed grips and dead load conditions. Since the load is kept constant during the extension, $\delta Q_i = 0$, and Equation 4 reduces to

$$\delta\Pi_{fixed\, Q_i} = -\int_{\Sigma_{cr}} 2G\delta l d\Sigma_{cr} \tag{7}$$

or

$$G_{fixed\ Q_i} = -\frac{\partial \Pi}{\partial S_{cr}} \tag{8}$$

which for the plate shown in Figure 2 reduces to $G_{fixed\ Q} = -\frac{1}{2b}\frac{\partial \Pi}{\partial \ell}$. Thus when the load is kept constant the release of strain energy and the work of the applied force both contribute to overcoming the energy required to grow the crack. When the material is linear elastic, the potential energy is the negative of the strain energy. For this case Equation 8 reduces to

$$G_{fixed\ Q_i} = \frac{\partial U}{\partial S_{cr}} \tag{9}$$

The next section explains how the ERR can be used to determine conditions for crack extension.

2.2 The generalized Griffith criterion

Griffith's pioneering work on fracture Griffith (1921) was concerned with the strength of glass, which for all intents and purposes can be assumed to be perfectly brittle. This means that the only resistance to crack propagation is provided by the energy required to create the additional top and bottom surfaces of the crack, and that it does not depend on the amount of crack growth that has occurred,. This material resistance is defined by the energy per unit crack surface area, 2γ. Griffith considered an isolated crack in a thin plate loaded by uniform far-field tension, for which he knew the functional form of the ERR at fixed load. He then stated that the equilibrium between the crack length and the applied stress is determined by the energy balance equation represented by equating the ERR to the material resistance

$$G = \frac{(\kappa+1)\,\sigma^2\pi a}{8\mu} = 2\gamma \tag{10}$$

where μ is the shear modulus, $\kappa = 3 - 4\nu$ for plane strain and $\kappa = (3 - \nu)/(1+\nu)$ for plane stress. The ERR always involves the square of a reference stress, the elastic moduli, and a characteristic dimension of the cracked component such as the crack length. The Griffith criterion represents an equilibrium between these parameters, and it can be used in different ways. For a structure containing a crack of given length, it determines the applied stress required to extend the crack. Similarly, the criterion can be used to determine the allowable crack length for a prescribed operating applied stress. Note that because its ERR is an increasing function of crack length and the resistance to crack propagation is constant, then under dead load conditions the Griffith crack grows in an unstable manner.

The Griffith Criterion was extended by Orowan (Orowan (1955)) to cracks whose growth involves additional types of energy dissipation occurring within a relatively small volume near their front and that are orders of magnitude larger than the surface energy. These include plastic deformation in metals, microcracking in concrete and ceramics, and crazing in polymers. As long as the extent of the "process zone" within which the dissipation occurs is sufficietly smaller than the crack length and other relevant characteristic lengths, then the elastic solution provides an accurate description of the strain intensity near the crack front, and the elastic ERR is meaningful. If these "small scale yielding" conditions prevail, then Orowan suggested that the equilibrium can be determined by treating the additional energy dissipation as an equivalent surface energy, $2\gamma^*$, so that the equilibrium condition becomes

$$G = 2\gamma + 2\gamma^* = \Gamma \tag{11}$$

There are numerous ways of introducing Equations 10 and 11 into linear elastic fracture mechanics. Here is but another. Equation 2 can be arranged to highlight the incremental work of the generalized force during crack extension

$$Q_i \delta q_i = \delta U + \int_{\Sigma_{cr}} 2G\delta l d\Sigma_{cr} \tag{12}$$

But as the crack extends, a portion of this additional work is transformed into stored elastic energy and the rest is dissipated to achieve the additional crack surfaces

$$Q_i \delta q_i = \delta U + \Gamma \delta S_{cr} \tag{13}$$

With the understanding that the equivalent surface energy has beeen assumed to be evenly distributed along the new crack surfaces, then a comparison of Equations 12 and 13 provides the crack extension condition

$$G = \Gamma \tag{14}$$

Table 1 lists the range of Γ for a variety of materials relevant to engineering (Ashby (1992)).

Table 1. Toughness values for different classes of materials (data taken from Ashby (1992)).

Material	Γ (kJ/m^2)
Metals	1-100
Wood	1-10
Polymers and elastomers	0.1-10
Ceramics	0.001-0.1

In summary, the energy approach to fracture involves determining the ERR for a given geometry and loading, and equating it with the material's resistance.

Example 1: Steady state interface crack propagating along a fiber/matrix interface. A beautiful example of how Equation 9 can be used to develop simplified analytical microstructural models of complex structures containing cracks involves the micromechanical (unit cell) model of a long fiber embedded in a matrix shown in Figure 3.

The fiber is perfectly bonded to the matrix everywhere except along the portion represented by the interface crack of length l. Denoting fiber and matrix with subscripts f and m, respectively, the elastic constants are E_f, ν_f and E_m, ν_m. The top surface of the cell represents the far field, where the fiber and matrix share the applied loading and experience a state of constant stress defined by σ_f and σ_m. The bottom surface of the matrix and the vertical external surface of the unit cell are both traction-free, while the fiber is subjected to a uniform traction, t. The problem consists of estimating, through a "back of the envelope" strength of materials analysis, the dependence of the interface crack's ERR on the elastic mismatch, $\Sigma \equiv \frac{E_m}{E_f}$, and fiber volume ratio, $f \equiv \frac{V_f}{V_m}$.

The problem was inspired by the observed effects of interface toughness on the ductility of composites comprised of brittle matrices and relatively long and brittle fibers (brittle-matrix-composites, or BMCs). Note that the fiber can perhaps be interpreted as an osteon surrounded by a homogeneized tissue within a bone structure. If the fiber/matrix interface in BMCs is sufficiently tough, then the matrix cracks that inevitably initiate and propagate in a direction perpendicular to the fibers and the applied loading will continue to propagate along that direction, fracturing the fibers they encounter. This will produce an unzipping of the composite and a load-displacement

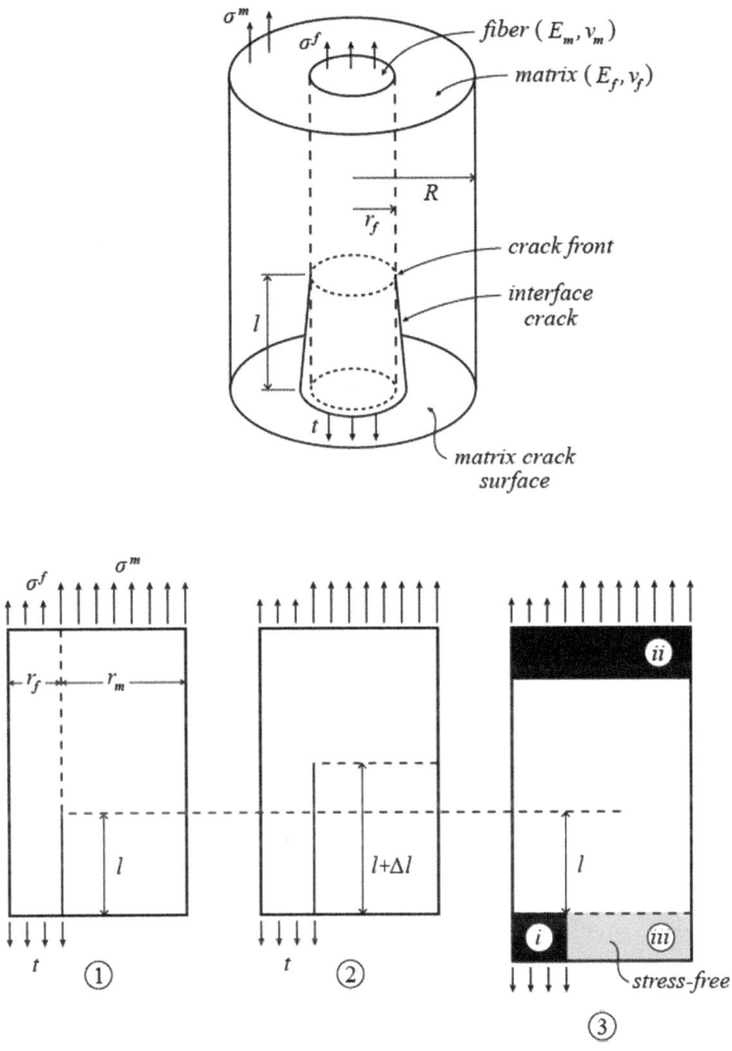

Figure 3. Unit cell model of fiber/matrix interface crack in unidirectionally reinforced composite (adapted from Charalambides and Evans (1989)).

curve that will exhibit a catastrophic post-peak behavior, indicating relatively low toughness. However if the toughness of the interface is sufficiently low compared to the toughness of the fibers, then as shown in Figure 4the matrix cracks will be deflected along the fiber/matrix interfaces. For this case an increase in load is required to initiate and extend additional matrix cracks, which in turn will be

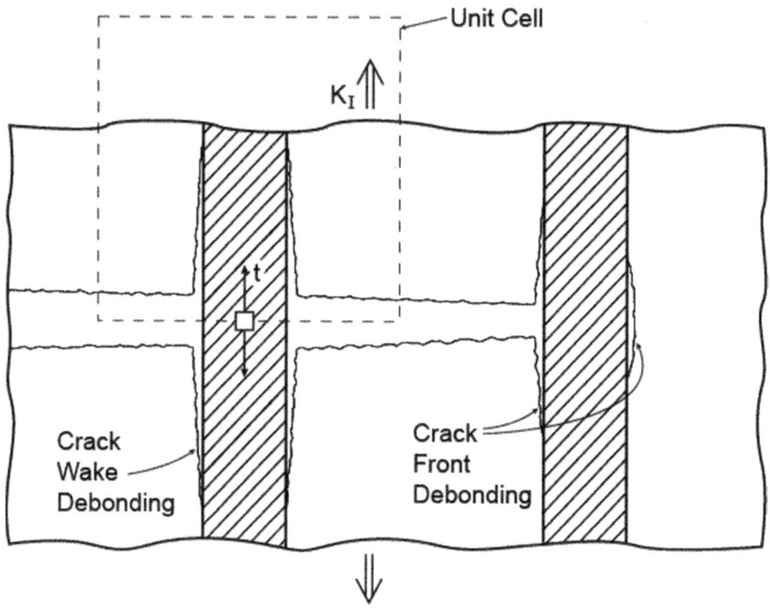

Figure 4. Matrix cracks deflected along the fiber/matrix interfaces in a brittle matric composite.

deflected along the interfaces they will encounter. The process will repeat itself until the matrix cracks extend across the whole specimen while the fibers remain intact. Eventually the fibers will rupture. However the load-displacement curve will exhibits a graceful failure and a much larger amount of dissipated energy. This scenario has been been observed in experiments, as reported by Evans (Evans (1989)). An estimate of the ERR of the fiber/matrix interface crack can shed light on the roles played by the elastic mismatch, fiber volume fraction, and fiber to matrix toughness ratio in determining which scenario will be realized. This is because in order for the cracks to be deflected along the interface, it is necessary that the

ERR of the interface crack be higher than the toughness of the interface, and that the ERR of a crack that could potentially penetrate the fibers must be smaller than the toughness of the fibers. Charalambides and Evans (Charalambides and Evans (1989)) proposed the unit cell model described above to estimate the dependence of the ERR on the parameters that define the composite material; the elastic mismatch between the fibers and the matrix, and the volume fraction of fibers. Knowing this dependence will guide the materials engineer to design the composite structure and the toughness of the interface and the fibers in such a way that the scenario will be realized. The choice of the micromechanical model emphasizes the importance of reducing complex problems to simplified models that retain the essential features of the phenomena and that are amenable to analytical treatment. In this application the original problem sketched in Figure 4 is a difficult boundary value problem involving a main crack interacting with numerous fibers and multiple matrix and interface cracks. In principle one can solve such a specific boundary value problem, but its solution will not shed much light on the role played by the system's control variables. The unit cell model, in contrast, retains the essential features of Figure 4. In addition to involving the elastic mismatch, the fiber volume fraction, and the fiber/matrix interface crack, the micromechanical model described above approximates the matrix crack through the enforcement of zero traction along the bottom surface of the matrix, and realizes composite behavior through the load-sharing prescribed along the top surface of the unit cell. Figure 3 includes three planar views of a sector of the unit cell. Going from left to right, the first view shows the initial interface crack, and the second and third views represent the configuration corresponding to an extension Δl. The third view indicates that as long as the crack is sufficiently long (for this case it was determined by solving the boundary value problem that the crack should be longer than the diameter of the fiber) so that the state of stress in region (i) is uniform tension, then the strain energies of the configurations prior and after crack extension differ only by the those indicated in the shaded regions (i), (ii) and (iii). Noting that region (iii) is stress free, then it is clear that the final state can be rendered energy equivalent to the initial state by adding to it the strain energy in region (ii) and subtracting from it the stored energy in region (i). This means that the ERR of the interface crack is independent of the crack length, a situation that is referred to by the fracture mechanics community as steady-state cracking. Note that if the crack is short, then the ERR will depend on the crack length because the crack front has a stronger interaction with the free surface and the stress fields within regions (i) and (iii) are not uniform. This insight provides the following relation for the variation of strain energy that results from the additional crack surface area

$$\delta U = U_i - U_{ii} = G_{ss}(2\pi r_f)\Delta l \tag{15}$$

where r_f represents the fiber radius and G_{ss} is the steady state ERR.

Because the stress in region (i) is uniform, its strain energy is readily determined as $U_i = \frac{1}{2}tV_i\varepsilon_i = \frac{1}{2E_f}t^2\pi r_f^2\Delta\ell$. In region (ii) the stresses are also uniform, and they can be determined using equilibrium and strain compatibility; $tf = f\sigma_f + (1-f)\sigma_m$ and $\varepsilon_f = \varepsilon_m = \frac{\sigma_f}{E_f} = \frac{\sigma_m}{E_m}$. Solution of these equations leads to $U_{ii} = \frac{\pi r_f^2\Delta\ell}{2E_f}\left\{\left(\frac{\sigma_f}{t}\right)^2 + \left(\frac{r_m}{r_f}\right)^2\frac{1}{\Sigma}\left(\frac{\sigma_m}{t}\right)^2\right\}t^2$, where r_m represents the thickness of the annular matrix. Substitution of the strain energies into Equation 15 gives

$$\frac{4G_{ss}E_f}{t^2 r_f} = \frac{\Sigma}{\Sigma + \frac{f}{1-f}} \tag{16}$$

Figure 5 presents plots of Equation 16 in solid lines and selected results obtained using a finite element method analysis for $\nu_f = \nu_m = 0.25$. It is observed that this back of the envelope procedure is remarkably accurate for this steady-state configuration. The results provide valuable information that can guide the "material by design" paradigm.

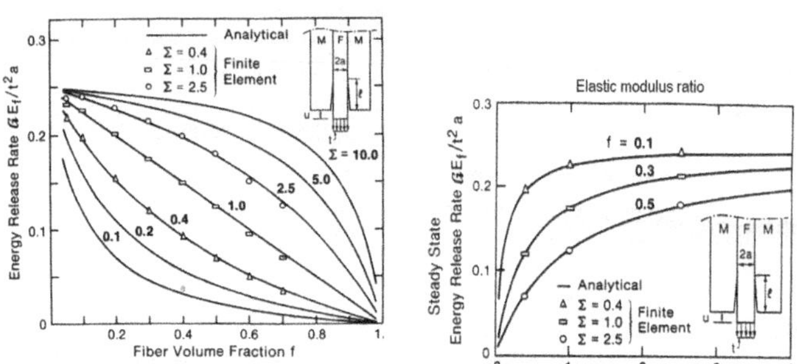

Figure 5. Steady-state energy release rate of a crack at the interface of a fiber embedded within a matrix (adapted from Charalambides and Evans (1989)).

For example, the plot of ERR as functions of fiber volume fraction suggests that if the fibers are stiffer than the matrix then the ERRs are strong decreasing functions of volume fraction. Therefore high volume fractions of such fibers may suppress interface cracking. The ERRs associated with fibers that are much more compliant than the matrix are much larger, weak functions of volume fraction, and can increase the desired propensity towards interface cracking.

3 Relationship between ERR, crack-tip fields and structural compliance

3.1 The Irwin formula

This section shows how the ERR, which represents a global measure, is directly related to the stress and strain intensities in the immediate vicinity of the crack front as follows. Consider a crack that extends along its defining plane; non-planar extension can be treated in a similar manner but will result in equations involving the direction of propagation. As shown in Appendix A, the variation in potential energy that results from crack extension can be written as

$$\delta\Pi = \delta U = -\frac{1}{2}\int_{\Sigma_{cr}} d\Sigma_{cr} \int_0^{\delta l} \sigma_{2i}^b (u_i^+ - u_i^-)^a dr \tag{17}$$

where r measures the perpendicular distance from a point ahead of the crack (in the 1-3-plane of Figure 1) to the crack front, σ_{2i}^b represents the traction components that existed across the plane that defines the extension (before the extension), and $(u_i^+ - u_i^-)^a$ represent the crack opening and crack sliding displacements of the new surfaces (after the extension). Therefore

$$G = -\frac{\partial U}{\partial S_{cr}} = \underset{\delta S_{cr}\to 0}{Lim}\frac{1}{\delta S_{cr}}\frac{1}{2}\int_{\Sigma_{cr}} d\Sigma_{cr} \int_0^{\delta l} \sigma_{21}^b (u_i^+ - u_i^-)^a dr \tag{18}$$

For the infinitesimal extension involved in the definition of the ERR, it suffices to use in the integrands of Equation 18 the well-established forms of the stresses and displacements in the vicinity of the crack front. These can be written in terms of the stress intensity factors, K_I, K_{II}, K_{III} as

$$\left\{\begin{array}{c}\sigma_{22}^b\\ \sigma_{21}^b\\ \sigma_{23}^b\end{array}\right\} = \frac{1}{\sqrt{2\pi r}}\left\{\begin{array}{c}K_I\\ K_{II}\\ K_{III}\end{array}\right\} \tag{19}$$

$$\left\{\begin{array}{c}u_1^+ - u_1^-\\ u_2^+ - u_2^-\\ u_3^+ - u_3^-\end{array}\right\}^a = \frac{(\kappa+1)}{\mu}\sqrt{\frac{\delta\ell - r}{2\pi}}\left\{\begin{array}{c}K_{II}(\ell+\delta\ell)\\ K_I(\ell+\delta\ell)\\ 4K_{III}(\ell+\delta\ell)\end{array}\right\} \tag{20}$$

The stress intensity factors depend on the exact nature of the details of the structural geometry and loading conditions, and they can be determined for specific configurations through the solution of the elasticity boundary value problem. The mode-I factor K_I is associated with the symmetric opening of the crack surfaces, K_{II} is the mode-II factor associated with the in-plane asymmetric relative sliding of the crack surfaces, and K_{III} is the mode-III factor associated with the out-of-plane relative sliding of the crack surfaces. It is the purpose of the following discussion to show that the stress intensity factors are directly related to the ERR, and can therefore be estimated with excellent accuracy using simplified mechanical models. Note that in Equation 20 the stress intensity factors and the distance from the extended crack tip, $\delta l - r$, are those of the extended crack. However, in the limit $K_I(l + \delta l) \to K_I(l)$. Substitution of Equations 19 and 20 into Equation 18 provides the Irwin formula (Irwin (1957))

$$G = \frac{(\kappa + 1)}{8\mu}(K_I^2 + K_{II}^2 + K_{III}^2) \tag{21}$$

that links the ERR to the parameters that uniquely characterize the intensity of the stress and strain fields in the vicinity of the crack front. Note that the derivation of Equation 21 makes use of the identity $\int_0^{\delta l} \sqrt{\frac{\delta l - x}{x}}dx = \frac{\pi \delta l}{2}$.

Consider the mode-I crack problem. Equation 21 shows that for a crack growing straight the extension condition given by Equation 14 is equivalent to a critical stress intensity factor condition, $K_I = K_I^{critical} = \sqrt{\frac{8\mu\Gamma}{(\kappa+1)}}$. Therefore conditions for crack extension can be established with knowledge of the stress intensity factor. As shown next, the Irwin relation, when combined with either stress analysis or experiments, provides a very powerful tool for calculating not only the stress intensity factor, but also the stiffness of a cracked structure.

3.2 Relation between ERR and compliance

Knowledge of the ERR allows one to determine the reduction in stiffness, or alternatively the increase in compliance of the structure produced by the introduction of the crack as follows. Without loss of generality, consider the mode-I situation of a cracked plate of unit thickness subjected to a generalized point force. Equations 2 and 3 read

$$\delta U(q, \ell) = Q\delta q - 2G\delta\ell \tag{22}$$

$$\delta\Pi(Q, l) = -q\delta Q - 2G\delta l \tag{23}$$

from which it follows that

$$Q = \frac{\partial U(q,l)}{\partial q} \tag{24}$$

$$q = -\frac{\partial \Pi(Q,l)}{\partial Q} \tag{25}$$

$$2G = -\frac{\partial U(q,l)}{\partial l} = -\frac{\partial \Pi(Q,l)}{\partial l} \tag{26}$$

Equations 24-26 in turn provide the Maxwell relations

$$\frac{1}{2}\frac{\partial q(Q,l)}{\partial l} = \frac{\partial q}{\partial S_{cr}} = \frac{\partial G(Q,l)}{\partial Q} \tag{27}$$

$$\frac{1}{2}\frac{\partial Q(q,l)}{\partial l} = -\frac{\partial Q}{\partial S_{cr}} = -\frac{\partial G(q,l)}{\partial q} \tag{28}$$

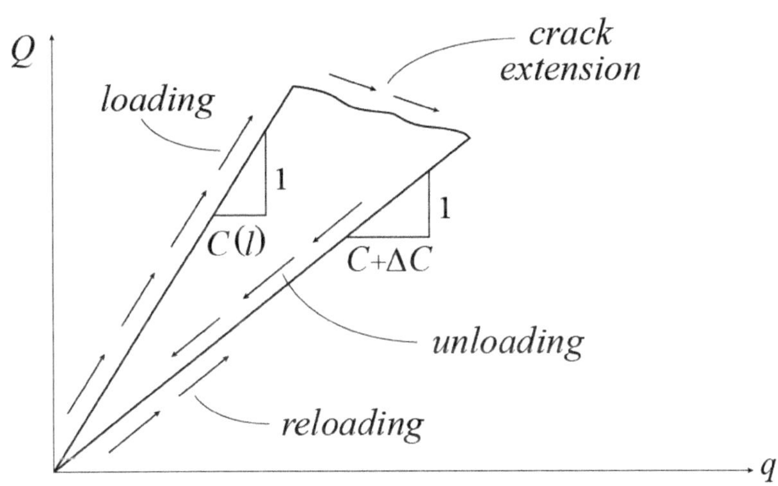

Figure 6. Change in compliance with respect to crack extension.

For a linear elastic solid the displacement and stress intensity factor are proportional to the load, and can therefore be expressed as

$$q = C(l)Q \tag{29}$$

$$K_I = Qk_I(l) \tag{30}$$

The compliance, C, and the unit force stress intensity factor function, k_I, have been written as functions of only crack length, but it is understood to also depend on material constants and structural geometry. The Irwin relation can be recast as

$$G = \frac{(\kappa+1)K_I^2}{8\mu} = \frac{(\kappa+1)}{8\mu}Q^2 k_I^2 \tag{31}$$

Differentiating Equations 29 and 31 with respect to crack length and load, respectively, and enforcing 27 gives

$$\left(\frac{\partial q}{\partial l}\right)_{fixed\,q} = Q\frac{dC(l)}{dl} = \left(\frac{\partial G}{\partial Q}\right)_{fixed\,l} = \frac{(\kappa+1)}{4\mu}K_I\left(\frac{\partial K_I}{\partial Q}\right)_{fixed\,l} = \frac{(\kappa+1)}{4\mu}Qk_I^2(l) \tag{32}$$

and finally

$$\frac{(\kappa+1)}{8\mu}K_I^2 = G = \frac{1}{2}Q^2\frac{dC(l)}{dl} \tag{33}$$

Equation 33 links the ERR and the stress intensity factor to the global compliance of the cracked component. It has numerous applications, including the following:

1. For a specimen with known stress intensity factor function and initial crack length, experimental or computational measurement of the change in compliance can be used to estimate the small amount of crack extension that has occurred. As illustrated in Figure 6, this involves increasing the load to a value that produces crack extension, unloading, reloading, meauring the change in compliance, and using the approximate formula

$$\Delta\ell = \frac{4\mu}{(\kappa+1)\,k_I^2}\Delta C \tag{34}$$

2. In a similar manner, for a given loading and crack configuration, k_I can be determined experimentally or computationally using Equations 31 and 30 to obtain

$$k_I = \frac{K_I}{Q} = 2\sqrt{\frac{\mu}{(\kappa+1)}}\sqrt{\frac{\Delta C}{\Delta l}} \tag{35}$$

3. For a given configuration, the stress intensity factor can be estimated through Equation 33 using a simple strength of materials analysis that can produce an accurate formula for the change of the compliance with respect to crack length.

4. The compliance of a cracked structure with known stress intensity factor can be determined by integrating Equation 33 .

The following examples illustrate applications 3 and 4.

Example 2: The ERR and stress intensity factor of the double cantilever beam specimen The classic example that is often used to demonstrate the use of Equation 33 to estimate the ERR and the stress intensity factor is the double cantilever beam (DCB) specimen shown in Figure 7.

Various designs of this popular experimental specimen are modifications of this generic trousers geometry. The loading consists of concentrated forces applied at the ends of the trouser. The specimen's name originated from the rigid-wall boundary conditions that were used in the simplest beam-theory treatment of the crack problem. As shown in Figure 7, the simplest model assumes that each leg can be represented by a beam with zero vertical displacement and zero slope along a plane that intersects the crack front perpendicularly. These highly idealized boundary conditions are expected to introduce, except for very slender beams, a substantial error in the compliance, the strain energy and in turn the ERR and the stress intensity factor. Elementary beam theory provides

$$\frac{u}{2} = \frac{4Pl^3}{EBh^3} \tag{36}$$

so that the compliance is

$$C = \frac{\frac{u}{2}}{P} = \frac{8l^3}{EBh^3} \tag{37}$$

and therefore

$$G_{fixedP} = G_{fixedu/2} = \frac{1}{2B}P^2\frac{dC}{dl} = \frac{12P^2l^2}{EB^2h^3} = \frac{3Eu^2h^3}{16l^4} \tag{38}$$

The Irwin formula provides the normalized stress intensity factor

$$\frac{K_I(l)h^{\frac{3}{2}}}{Pl} = \sqrt{12} \tag{39}$$

The idealized boundary conditions have resulted in a stress intensity factor formula that is accurate only for relatively long cracks, as shown in Figure 8 which compares Equation 39 with the (practically exact) formula derived by Gross and Srawley (Gross and Srawley (1966)) using the boundary collocation method

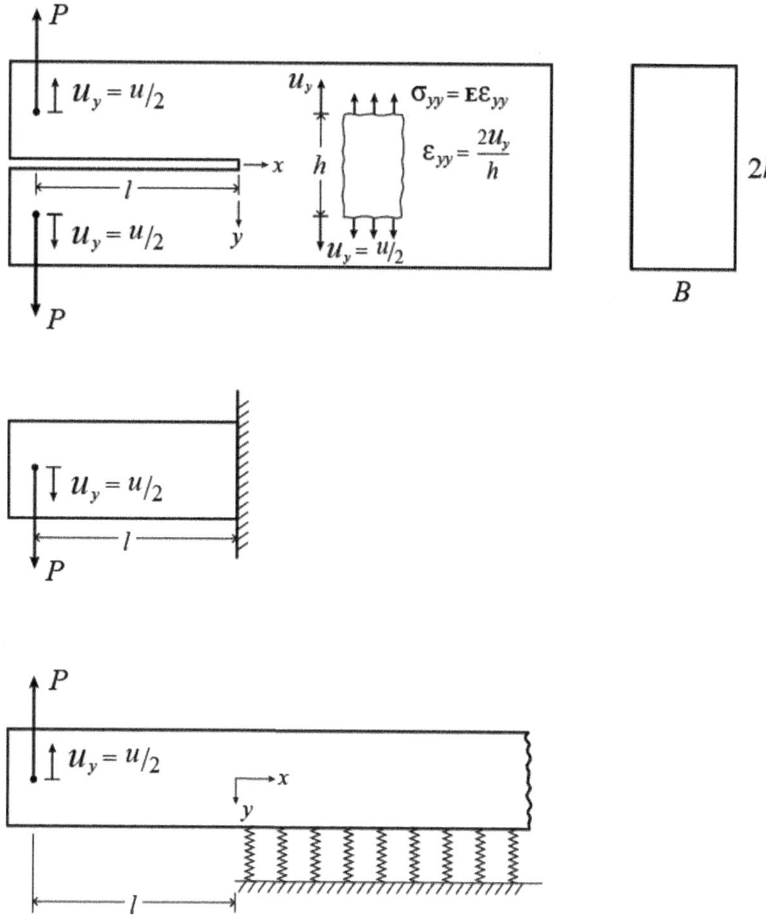

Figure 7. Double cantilever specimen and two beam theory approximations.

$$\frac{K_I(\ell)\,h^{3/2}B}{P\ell} = \sqrt{12}\left(1 + 0.69\frac{h}{\ell}\right) \tag{40}$$

A much improved formula can be derived by improving the model to account for the strain energy ahead of the crack, which involves abandoning the cantilever conditions adopted in the solution above. To this end the model shown in Figure 7 replaces the uncracked ligament by an elastic foundation comprised of a continuous distribution of springs with stiffness per unit width

$$k = \frac{2E}{h} \tag{41}$$

The spring stiffness is derived from the uniaxial constitutive relation of an element of gage length h far ahead of the crack tip, for which the displacement is equal to twice the center-line displacement of each beam, u_y. The strain and corresponding stress are equal to $\varepsilon_{yy} = \frac{2u_y}{h}$ and $\sigma_{yy} = E\varepsilon_{yy} = \frac{2u_y}{h}$, respectively.

The compliance of the beam is determied by dividing the beam into two parts: segments $-l \leq x \leq 0$, with corresponding deflection curve u_y^l, and $0 \leq x \leq \infty$, with corresponding deflection curve u_y^r. The governing equations are

$$EI\frac{d^4u_y^l}{dx^4} = q(x) = 0 \tag{42}$$

$$EI\frac{d^4u_y^r}{dx^4} = q(x) = -ku_y^r \tag{43}$$

The solutions of Equations 42 and43 are

$$u_y^l = c_1x^3 + c_2x^2 + c_3x + c_4 \tag{44}$$

$$u_y^r = e^{-\beta x}\left\{c_5 sin(\beta x) + c_6 cos(\beta x)\right\} + e^{\beta x}\left\{c_7 sin(\beta x) + c_8 cos(\beta x)\right\} \tag{45}$$

where $\beta \equiv \left(\frac{k}{EI}\right)^{\frac{1}{4}} = \frac{6^{\frac{1}{4}}}{h}$. In order for the displacement to be bounded at all points $c_7 = c_8 = 0$. The following conditions provide the remaining unknown constants of integration.

The left end of the beam has zero moment and a shear force that balances the applied force

$$EI\frac{d^2u_y^l(-l)}{dx^2} = 0 = -6c_1 l + 2c_2 \tag{46}$$

$$EI\frac{d^3u_y^l(-l)}{dx^3} = P = -6EIc_1 \tag{47}$$

Continuity of displacement, slope, moment and shear force at the junction between the two segments

$$u_y^l(0) = u_y^r(0) \Rightarrow c_4 = c_6 \tag{48}$$

$$\frac{du_y^l(0)}{dx} = \frac{du_y^r(0)}{dx} \Rightarrow c_3 = \beta(c_5 - c_6) \tag{49}$$

$$EI\frac{d^2u_y^l(0)}{dx^2} = EI\frac{d^2u_y^r(0)}{dx^2} \Rightarrow c_2 = -c_5\beta^2 \tag{50}$$

$$EI\frac{d^3u_y^l(0)}{dx^3} = EI\frac{d^3u_y^r(0)}{dx^3} \Rightarrow 3c_1 = 2\beta^3(c_5 + c_6) \tag{51}$$

Solution of the six equations above leads to

$$C = \frac{u}{P} = -\frac{2u_y^l(-l)}{P} = \frac{2}{EI}\left(\frac{l^3}{3} + \frac{l^2}{\beta} + \frac{1}{2\beta^3} + \frac{l}{\beta^2}\right) \tag{52}$$

which together with the definition of β provides

$$G = \frac{1}{2B}P^2\frac{dC}{dl} = \frac{12P^2}{Eh^3}\left(\frac{h^2}{6^{\frac{1}{2}}} + \frac{2lh}{6^{\frac{1}{4}}} + l^2\right) \tag{53}$$

The desired stress intensity factor follows from the Irwin formula

$$\frac{K_I\left(\ell\right)h^{\frac{3}{2}}B}{P\ell} = \sqrt{12}\left(1 + 0.64\frac{h}{\ell}\right) \tag{54}$$

As shown in Figure 8, Equations 40 and 54 are practically indistinguishable; the beam on the elastic foundation model does an excellent job of capturing the energy stored in the specimen. This example, together with the micromechanical model of the BMC discussed previously, drives home the point that the global compliance and the local stress intensity factor and ERR rate of a cracked component can be estimated with excellent accuracy with proper accounting of the strain energy dependence on crack length.

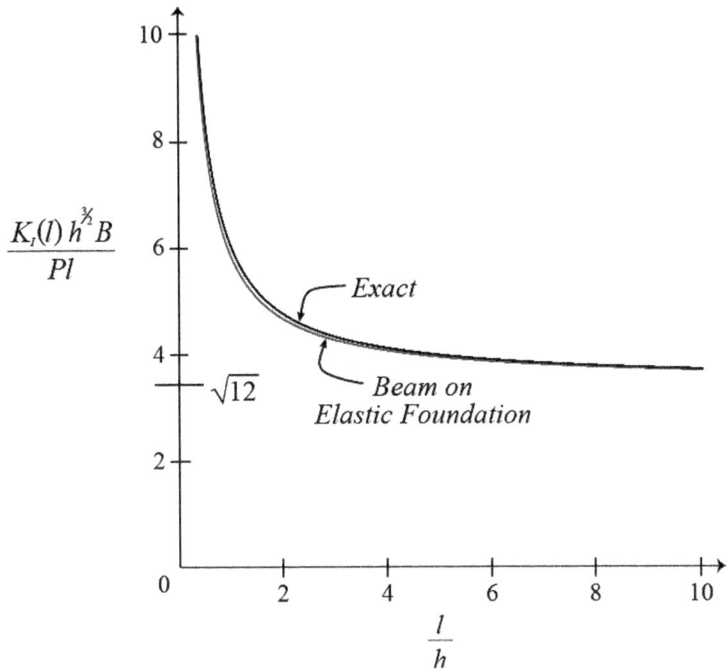

Figure 8. Stress intensity factor of double cantilever specimen.

3.3 The compliance of a cracked component

Much can be learned about the structural behavior of a cracked component with the knowledge of how its compliance is influenced by crack length and other characeristics of its geometry. The change in compliance with respect to an increase in crack length can be determined by integrating Equation 33 as follows

$$C(l_2) - C(l_1) = \frac{(\kappa + 1)}{4\mu Q^2} \int_{l_1}^{l_2} K_I^2(l)dl \tag{55}$$

Define the compliance of the uncracked configuration, $C^{nc} \equiv C(l_1 = 0)$, and the compliance contributed by the presence of the crack, $C^{cr} \equiv C(l_2 = l)$. Then Equation 55 allows the total compliance to be written as

$$C(l) = C^{nc} + C^{cr} = C^{nc} + \frac{(\kappa + 1)}{4\mu Q^2} \int_{l_1}^{l_2} K_I^2(l)dl \tag{56}$$

Equation 56 makes it very clear that in general the contribution to the reduction in stiffness cannot be determined by simply accounting for the reduction in cross-sectional area associated with the introduction of the crack.

Example 3: Structural response of an edge-cracked beam of varying slenderness. Handbooks are available that list stress intensity factor functions for a very large number of two and three-dimensional geometries. Equation 56 shows that these functions can be used to determine the compliance of a cracked structure in addition to the critical combination of applied stress and crack length that leads to crack extension. The Stress Analysis of Cracks Handbook (Tada et al. (1985)), referred to as Tada's Handbook, presents equations produced by the intergration of 56 for a large number of cracked components of practical interest. The configuration shown in Figure 9 is considered next as an example of how such equations could be used to learn about the structural response of a cracked component.

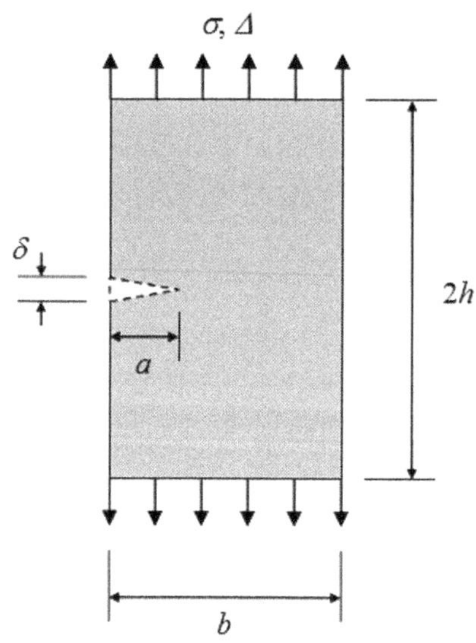

Figure 9. Extension of a planar crack in a three-dimensional solid

Tada's Handbook provides the following equations for the stress intensity factor

$$K_I = \sigma\sqrt{\pi a}F\left(\frac{a}{b}\right) \tag{57}$$

$$F\left(\frac{a}{b}\right)=\sqrt{\frac{2b}{\pi a}\tan\left(\frac{\pi a}{2b}\right)}\frac{0.752+2.02\frac{a}{b}+0.37\left[1-\sin\left(\frac{\pi a}{2b}\right)\right]^3}{\cos\left(\frac{\pi a}{2b}\right)} \tag{58}$$

As stated in the this handbook, Equations 57 and 58 were used in the integration that appears in Equation 56 to determine the crack's contribution to the load-point displacement. The results of this procedure were fitted by the following expression

$$\Delta^{cr}=C^{cr}P=\frac{4\sigma}{E'}V_2\left(\frac{a}{b}\right) \tag{59}$$

$$V_2\left(\frac{a}{b}\right)=\frac{a}{b}\frac{1}{\left(1-\frac{a}{b}\right)^2}\left[0.99-\frac{a}{b}\left(1-\frac{a}{b}\right)\left(1.3-1.2\frac{a}{b}+0.7\frac{a^2}{b^2}\right)\right] \tag{60}$$

For this simple geometry and loading the displacement in the absence of the crack is simply

$$\Delta^{nc}=2\frac{\sigma h}{E'} \tag{61}$$

so that the total displacement becomes

$$\Delta=2\frac{\sigma a}{E'}\left[2V_2\left(\frac{a}{b}\right)+\frac{h}{a}\right] \tag{62}$$

Knowledge of the compliance can provide useful insight into structural response, and in particular the design of experiments. Consider for example the problem of designing a fracture mechanics experiment using the geometry defined by Figure 9, and assume that the behavior of the specimen post-peak load is to be captured by running the experiment under displacement control. As will be shown next, this can be accomplished only if the aspect ratio of the specimen, $\frac{2h}{b}$, is chosen to be sufficiently small.

The load-displacement curve that would be realized under the condition that the crack extends at a critical value of the stress intensity factor, $K_I^{critical}$, can be constructed using the following procedure. The stress that is in equilibrium with a particular crack length is

$$\sigma_c=\frac{K_I^{critical}}{\sqrt{\pi a}F\left(\frac{a}{b}\right)} \tag{63}$$

A convenient reference strength that enables nondimensionalization of the load-displacement plots is the critical stress corresponding to vanishing small preexisting cracks, $\frac{a}{b}\rightarrow 0$, which is equal to

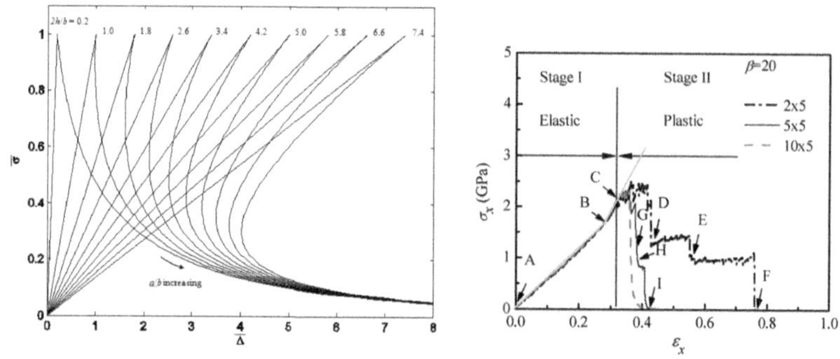

Figure 10. Normalized stress-displacement curve of edge-notched beam as functions of slenderness ratio (left); Stress-strain curves of collagen fibrils as functions of molecules along their lengths simulated using multiscale molecular dynamics model (right) (Tang et al. (2010)).

$$\sigma_c^o \equiv \frac{K_I^{critical}}{1.122\sqrt{\pi a}} \tag{64}$$

The normalized equilibrium stress is therefore equal to

$$\bar{\sigma} \equiv \frac{\sigma_c}{\sigma_c^o} = \frac{1.122}{F\left(\frac{a}{b}\right)} \tag{65}$$

and the nondimensional conjugate displacement can be defined by

$$\bar{\Delta} \equiv \frac{\Delta E'}{b\sigma_c^o} = 2\bar{\sigma}\frac{a}{b}\left[2V_2\left(\frac{a}{b}\right) + \frac{h}{a}\right] = \frac{2.244}{F\left(\frac{a}{b}\right)}\left[2\frac{a}{b}V_2\left(\frac{a}{b}\right) + \frac{h}{b}\right] \tag{66}$$

The load-displacement curve is shown in Figure 10 for selected values of the slenderness, $\frac{h}{b}$. It is observed that for displacement controlled experiments there exists a transition from stable to unstable post-peak response characterized by the snap-back instability. The responses illustrated in Figure 10 provide valuable guidance for the design of experiments on cracked specimens. Specifically, the results of this example indicate that displacement control could be used to capture the post-peak response only for beams with aspect ratios less than ~ 1.8. If the slenderness is greater than this amount, then under displacement control when the peak load is reached the load-displacement curve would drop down to the lower stable branch. The consequence of this is that if one was not aware of the specimen's intrinsic snap-back response, then the area under the curve traced by the discontinuous jump from peak load to the lower branch could be incorrectly and unconservatively interpreted as the fracture energy of the structure. In order to capture the snap-back instability during a computational simulation or an experiment, what is required is a control that dictates a constant rate of crack advance. This could be accomplished by controlling the crack mouth opening displacement (CMOD), δ, which is also given in the handbook and can be normalized as

$$\bar{\delta} = \frac{\delta E'}{b\sigma_c^o} = 4\bar{\sigma}\frac{a}{b}V_1\left(\frac{a}{b}\right) = \frac{4.488}{F\left(\frac{a}{b}\right)}\frac{a}{b}V_1\left(\frac{a}{b}\right) \tag{67}$$

$$V_1\left(\frac{a}{b}\right) = \frac{1.46 + 3.42\left[1 - cos\left(\frac{\pi a}{2b}\right)\right]}{cos^2\left(\frac{\pi a}{2b}\right)} \tag{68}$$

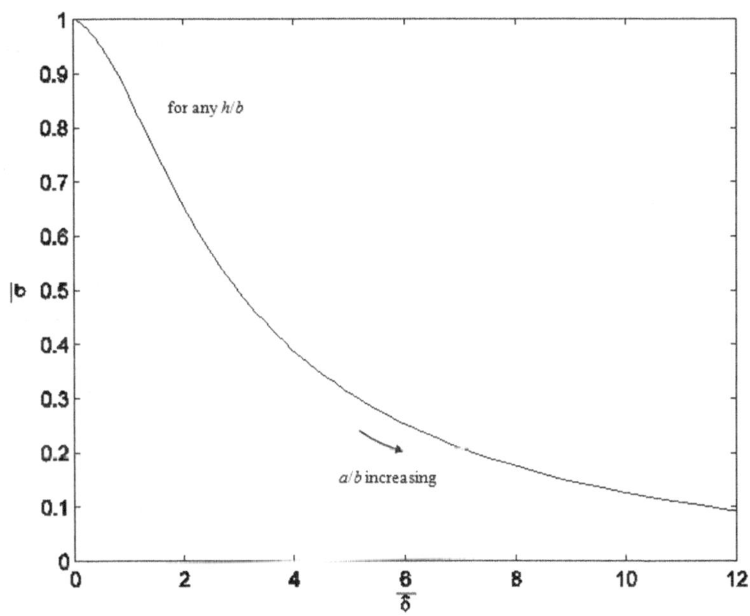

Figure 11. Normalized stress versus crack mouth opening displacement of edge-cracked beam.

which the plot in Figure 11 shows is independent of slenderness ratio and monotonic. Therefore performing an experiment using CMOD control is equivalent to controlling the rate of crack extension. What is remarkable is that Griffith equilibrium demands that during the snap-back response the load does negative work.

The behavior of the edge-notched beam explains the qualitative behavior of collagen fibrils subjected to uniaxial tension presented in Chapter 3. Specifically, the experiments demonstrated that the fibrils, which have very large aspect ratios, fail in a brittle manner indicated by a lack of post-peak response under displacement control. This is consistent with the results shown in Figure 10, as long as the flaws that are created within the fibrils can be idealized as a crack in a very slender rod.

The simplified continuum model could be used to assess the qualitative predictions of multiscale molecular dynamics based models, including the one presented in Tang et al. (2010). In this study simple elongation of collagen fibrils consisting of m collagen molecules along the fibril length and n molecules across the thickness were simulated for different levels of cross-linking density, β. Representative results shown in Figure 10 for n=2, m=2,5,10 and $\beta = 20$, are in qualitative agreement with the continuum model. For this prescribed density of cross-links, the fibril comprised of two molecules along its length exhibits stability under displacement control as evidenced by the gradual reduction in load in the post-peak region, while the one made of ten molecules along its length snaps into two pieces once the peak load is reached. The molecular dynamics simulations are performed using velocity control, and therefore the snap-back instability that is inherent in the slender fibrils is not captured.

4 Summary

This chapter has demonstrated that simplified continuum mechanics models could be often be used to study the mechanical response of complex multiscale structures, and to explain the results of simulations performed using multiscale computational models.

Appendix A

Consider two loadings on the extended crack as shown in Figure 12. The figure on the right highlights the crack opening displacements of the traction-free extension, $\left(u_i^+ - u_i^-\right)^a$. In the configuration on the left the crack surfaces along the extension are kept from displacing relative to each other by the tractions that existed along the extension prior to extension, $t_2^b = \sigma_{2i}^b n_i$. Denote the fields before and after crack extension by subscripts b and a, respectively, so that for any quantity t, $\delta t \equiv t^a - t^b$. If the loaded extension in the figure on the left is treated as an additional surface with prescribed traction, then the corresponding virtual work equation for arbitrary kinematically admissible variations of displacements is

$$\int_V \sigma^b_{ij}\delta\varepsilon^k_{ij}dV = \int_{S_t} t^o_i\delta u^k_i dS_t + \int_{\delta S_{cr}} t^b_2\delta u^k_i d\delta S_{cr} + \int_V f_i\delta u^k_i dV \tag{69}$$

In the figure on the right the extension is traction-free, and the virtual work equation is

$$\int_V \sigma^a_{ij}\delta\varepsilon^k_{ij}dV = \int_{S_t} t^o_i\delta u^k_i dS_t + \int_V f_i\delta u^k_i dV \tag{70}$$

Prescribe the kinematically admissible variations in displacements in Equation 69 as the displacements associated with the figure on the right, and the kinematically admissible variations in Equation 70 as the displacements associated with the figure on the left. Then

$$\int_V \sigma^b_{ij}\varepsilon^a_{ij}dV = \int_{S_t} t^o_i u^a_i dS_t + \int_{\delta S_{cr}} t^b_2 u^a_i d\delta S_{cr} + \int_V f_i u^a_i dV \tag{71}$$

$$\int_V \sigma^a_{ij}\varepsilon^b_{ij}dV = \int_{S_t} t^o_i u^b_i dS_t + \int_V f_i u^b_i dV \tag{72}$$

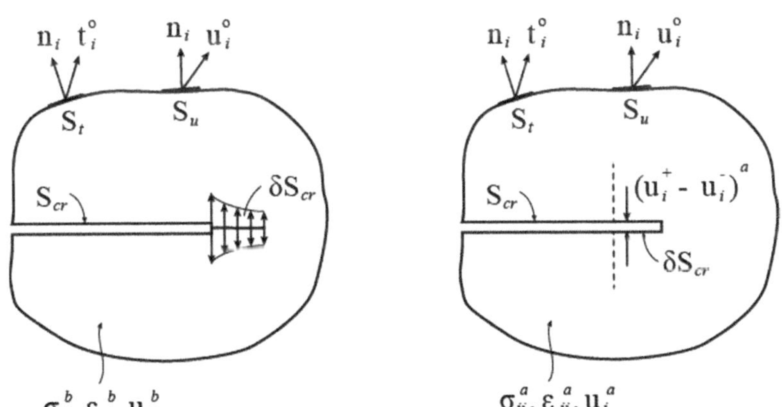

Figure 12. Crack extension subjected to traction that existed prior to extension (left) and traction-free extension (right).

According to Betti's reciprocal theorem (Sokolnikoff (1956)) the left-hand sides of Equations 71 and 72 are equal. Thus

$$\int_{S_t} t_i^o \left(u_i^a - u_i^b\right) dS_t + \int_V f_i \left(u_i^a - u_i^b\right) dV + \int_{\delta S_{cr}} t_2^b u_i^a d\delta S_{cr} = 0 \qquad (73)$$

or

$$-\int_{S_t} t_i^o \delta u_i dS_t - \int_V f_i \delta u_i dV = \int_{\delta S_{cr}} t_2^b u_i^a d\delta S_{cr} \qquad (74)$$

It is easy to show that the left-hand side of Equation 74 is equal to twice the change in potential energy resulting from crack extension. To this end consider that the potential energy of the traction-free crack before extension (figure on left with $\delta S_{cr} = 0$) is given by

$$\Pi^b = \frac{1}{2}\int_V \sigma_{ij}^b \varepsilon_{ij}^b dV - \int_{S_t} t_i^o u_i^b dS_t - \int_V f_i u_i^b dV \qquad (75)$$

The variation in potential energy for an arbitrary kinematically admissible displacement field is given by

$$\delta\Pi^b = \int_V \sigma_{ij}^b \delta\varepsilon_{ij}^k dV - \int_{S_t} t_i^o \delta u_i^k dS_t - \int_V f_i \delta u_i^k dV \qquad (76)$$

If the kinematically admissible variation is set equal to the actual field before extension, the potential energy should be minimum. Setting its variation equal to zero, its minimum value can be calculated as

$$min\Pi^b = -\frac{1}{2}\int_{S_t} t_i^o u_i^b dS_t - \frac{1}{2}\int_V f_i u_i^b dV \qquad (77)$$

Similarly, for the extended crack

$$min\Pi^a = -\frac{1}{2}\int_{S_t} t_i^o u_i^a dS_t - \frac{1}{2}\int_V f_i u_i^a dV \qquad (78)$$

Therefore the change in potential energy is given by

$$\delta\Pi = \Pi^a - \Pi^b = -\frac{1}{2}\int_{S_t} t_i^o \delta u_i dS_t - \frac{1}{2}\int_V f_i \delta u_i dV \qquad (79)$$

Substituting this last equation into Equation 74 provides

$$\delta\Pi = -\frac{1}{2}\int_{\delta S_{cr}} t_i^b u_i^a d\delta S_{cr} = -\frac{1}{2}\int_{\Sigma_{cr}} d\Sigma_{cr} \int_0^{\delta l} \sigma_{2i}^b (u_i^+ - u_i^-)^a dr \qquad (80)$$

Bibliography

M.F. Ashby. *Materials selection in mechanical design.* Pergamon Press, 1992.

P.G. Charalambides and A.G. Evans. Debonding properties of residually stressed brittle-matrix-composites. *Journal of the American Ceramic Society*, 70(5):746–753, 1989.

A.G. Evans. The mechanical performance of fiber-reinforced ceramic matrix composites. *Materials Science and Engineering*, A107:2167–2181, 1989.

A.A. Griffith. The phenomena of rupture and flow in solids. *Philosophical Trans. Royal Society, London*, A221:163–198, 1921.

B. Gross and J.E. Srawley. Stress-intensity factors by boundary collocation for single-edge notched specimens subjected to splitting forces. *NASA TND*, 3295, 1966.

G.R. Irwin. Analysis of stresses and strains near the end of a crack traversing a plate. *Journal of Applied Mechanics*, 64:361–364, 1957.

E. Orowan. Energy of fracture. *Welding Journal, Research Supplement*, 34(3):157s–160s, 1955.

I.S. Sokolnikoff. *Mathematical theory of elasticity.* McGraw-Hill Book Co., Inc., 1956.

H. Tada, P.C. Paris, and G.R. Irwin. *The stress analysis of cracks handbook, 2nd ed.* Del Research Corp., 1985.

Y. Tang, R. Ballarini, M.J. Buehler, and S.J. Eppell. Deformation micromechanisms of collagen fibrils under uniaxial tension. *Journal of the Royal Society Interface*, 7:839–850, 2010.

GPSR Compliance

The European Union's (EU) General Product Safety Regulation (GPSR) is a set of rules that requires consumer products to be safe and our obligations to ensure this.

If you have any concerns about our products, you can contact us on ProductSafety@springernature.com

In case Publisher is established outside the EU, the EU authorized representative is:

Springer Nature Customer Service Center GmbH
Europaplatz 3
69115 Heidelberg, Germany

Batch number: 10578109

Printed by Printforce, the Netherlands